AS UNIT 2

STUDENT GUIDE

CCEA

Biology

Organisms and biodiversity

John Campton

HODDER
EDUCATION
AN HACHETTE UK COMPANY

Hodder Education, an Hachette UK company, Blenheim Court, George Street, Banbury, Oxfordshire OX16 5BH

Orders

Bookpoint Ltd, 130 Park Drive, Milton Park, Abingdon, Oxfordshire OX14 4SE

tel: 01235 827827

fax: 01235 400401

e-mail: education@bookpoint.co.uk

Lines are open 9.00 a.m.–5.00 p.m., Monday to Saturday, with a 24-hour message answering service. You can also order through the Hodder Education website: www.hoddereducation.co.uk

This guide has been written specifically to support students preparing for the CCEA AS and A-level Biology examinations. The content has been neither approved nor endorsed by CCEA and remains the sole responsibility of the author.

Cover photo: andamanse/Fotolia; p. 69 Herve Conge, ISM/Science Photo Library

Typeset by Integra Software Services Pvt. Ltd, Pondicherry, India

Printed in Italy

Hachette UK's policy is to use papers that are natural, renewable and recyclable products and made from wood grown in sustainable forests. The logging and manufacturing processes are expected to conform to the environmental regulations of the country of origin.

Contents

Content Guidance

Questions & Answers

Section A Structured questions

Section B Essay questions

▉Getting the most from this book

Exam tips

Advice on key points in the text to help you learn and recall content, avoid pitfalls, and polish your exam technique in order to boost your grade.

Knowledge check

Rapid-fire questions throughout the Content Guidance section to check your understanding.

Knowledge check answers

1 Turn to the back of the book for the Knowledge check answers.

Summaries

- Each core topic is rounded off by a bullet-list summary for quick-check reference of what you need to know.

Exam-style questions

Commentary on the questions

Tips on what you need to do to gain full marks, indicated by the icon **e**

Section B Essay questions

▉Section B Essay questions

Question 18 The movement of water through the plant
Give an account of the processes involved in the movement of water through a plant. (15 marks)
In this question you will be assessed on your written communication skills.

Total: 15 marks

e In answering this question (worth 15 marks) you must write in continuous prose because your quality of written communication (QWC) will be assessed. This is a big topic, so take some time to *devise a plan*. Be aware that 'the movement of water through a plant' involves what happens in the root, in the stem and in the leaf. You must be able to sequence your ideas and use the appropriate biological terms, which is how QWC will be assessed.

Student A
Water enters the root and moves through the cortex via the apoplast or symplast pathways. ✓ Using the apoplast route, water moves through the cellulose cell walls. ✓ This is the main route ✓ since there is less resistance to movement. Water moves through the cytoplasm of cells and from cell to cell via plasmodesmata ✓ using the symplast route ✓. Water can only move through the endodermis, and into the xylem, using the symplast route. ✓

Forces of adhesion between the water molecules and the lignified xylem vessels ✓ help the water to creep up the xylem ✓. The force of cohesion, due to the attraction between water molecules, maintains a continuous water column. ✓ This water column moves up as water leaves the xylem to replace ✓ the water that has evaporated from the mesophyll surface ✓ and diffused out of the open stomata ✓. Throughout, water is moving along a water potential gradient ✓ as transpiration creates a particularly negative water potential in the leaf ✓ ▉

Quality of written communication ▉

e 15/15 marks awarded ▉ Student A has 14 correct points in the content. ▉ This is a well-structured account and ideas are expressed fluently. The links within the overall process are made clearly, all points are sequenced and biological terms are used accurately. Since the quality of written communication is of a very high standard, a total of 15 marks is judged to be appropriate.

Student B
In the root, water travels from cell to cell via two routes: the symplast and the apoplast. ✓ Apoplast is when the water travels through each cell ✗, whereas symplast is when water travels along the cell walls. ✗ However, water cannot pass through the Casparian strip ✓ and so must enter the cell before passing into the xylem. ✓ Water is then drawn upwards by negative tension, adhesion and cohesion. Negative tension is the force produced by water evaporating out of the leaves. ✗ Adhesion is when the polar water molecules ✓ are attracted to the sides of the xylem ✓. Cohesion is when water molecules are attracted to each

Sample student answers

Practise the questions, then look at the student answers that follow.

Commentary on sample student answers

Read the comments (preceded by the icon **e**) showing how many marks each answer would be awarded in the exam and exactly where marks are gained or lost.

■ About this book

The aim of this book is to help you prepare for the AS Unit 2 examination for CCEA Biology. It also offers support to students studying A2 biology, since topics at A2 rely on an understanding of AS material.

The **Content Guidance** contains everything that you should learn to cover the specification content of AS Unit 2. It should be used as a study aid as you meet each topic, for end-of-topic tests, and during your final revision. For each topic there are *exam tips* and *knowledge checks* in the margins. Answers to the knowledge checks are provided towards the end of the book. At the end of each topic there is a list of the practical work with which you are expected to be familiar. This is followed by a comprehensive, yet succinct, summary of the points covered in each topic.

The **Questions & Answers** section contains questions on each topic. There are answers written by two students with comments on their performances and how they might have been improved. These will be particularly useful during your final revision. There is a range of question styles to reflect those you will encounter in the AS Unit 2 exam, and the students' answers and comments will help with your examination technique.

Developing your understanding

It is important that through your AS course you develop effective study techniques.

- You must *not* simply read through the content of this book.
- Your understanding will be better if you are *active* in your learning. For example, you could take the information given in this book and present it in different ways:
  - a series of bullet points summarising the key points
  - an annotated diagram to show structure and function (e.g. a diagram of the heart with labelled features and brief notes on function) or an annotated graph (e.g. showing pressure changes during a cardiac cycle, with notes explaining the phases and points at which valves are opened and closed)
  - a spider diagram, e.g. one on the heart to include reference to its structure, the wave of excitation, the phases of the cardiac cycle, pressure changes and the operation of valves
- Compile a glossary of terms for each topic. Key terms in this guide are shown in **bold** (with some defined in the margin) and for each you should be able to provide a definition. This will develop your understanding of the language used in biology and help you where *quality of written communication* is being assessed.
- Write essays on different topics. For example, an essay on transport in plants will test your understanding of the entire topic and give you practice for the Section B question.
- *Think* about the information in this book so that you can *apply* your understanding in unfamiliar situations. Ultimately you will need to be able to deal with questions that set a topic in a new context.
- Ensure that you are familiar with the expected practical skills, as questions on these may be included in this unit.
- Use past questions and other exercises to develop all the skills that examiners must test.
- Use the topic summaries to check that you have covered all the material you need to know and as a brief survey of each topic.

Content Guidance

■ Principles of exchange and transport

Living cells require certain substances from their environment to maintain their metabolic processes, and need to remove the toxic by-products of metabolism.

Animal tissues obtain:

- oxygen from the air (or from water if they are aquatic)
- glucose, fatty acids and amino acids from ingested food
- water

Animal tissues remove carbon dioxide and nitrogenous waste (e.g. urea).

Plant tissues obtain:

- oxygen from the air, especially at night
- carbon dioxide from the air during the day
- inorganic ions (e.g. nitrate ions to provide nitrogen for amino acid synthesis and phosphate for the synthesis of phospholipids) from the soil solution
- water

Plant tissues remove either carbon dioxide or oxygen, depending on the time of day.

Any substances exchanged with the environment may need to be transported within the organism.

The exchange of substances

There are a number of factors that influence the absorption or exchange of substances.

The exchange of substances occurs only at *moist, permeable surfaces*. Some organisms have moist and permeable body surfaces (e.g. earthworm and frog). However, to prevent water loss by evaporation, mammals and flowering plants, which require terrestrial adaptations, possess impermeable surfaces. This means that mammals and flowering plants must have specialised absorptive or exchange surfaces.

The need for specialised absorptive or exchange surfaces depends on the size and shape of the organism, because both these factors affect an organism's **surface area-to-volume ratio**. The surface area is represented by the total number of cells in direct contact with the surrounding environment. The volume is the total three-dimensional space occupied by metabolically active tissues. The absorptive surface area is a measure of the rate of supply of metabolites to tissues. The volume of the organism is a measure of its demand for metabolites. An organism must be capable of taking up sufficient material to satisfy its needs. Therefore, the surface area-to-volume ratio is critical — it must be sufficiently large.

Exam tip

An **absorptive surface** allows the uptake of soluble substances — for example, products of digestion in the ileum or water and ions by root hairs. A **gas exchange surface** allows oxygen to move through in one direction and carbon dioxide in the opposite direction. It is sometimes referred to as a **respiratory surface** because the respiratory gases, oxygen and carbon dioxide, are involved.

Knowledge check 1

Which inorganic ion do plants need to absorb to make amino acids?

The influence of size on surface area-to-volume ratio is shown graphically in Figure 1, in which cubes of different dimensions are used as the example.

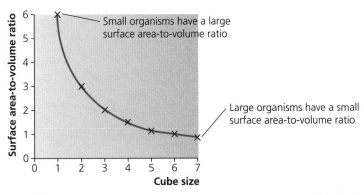

Figure 1 The relationship between size and surface area-to-volume ratio

While this relationship has been calculated for cubes, the principle is true for all regular shapes. *Small* organisms have a *large* surface area-to-volume ratio. The external surface of a small organism can be used as a **gas exchange** surface because the relatively large surface area is able to supply sufficient oxygen to the relatively small volume. For example, an earthworm is small enough for its body surface to be used for gas exchange.

A larger organism has a small surface area compared with its large volume, i.e. a small surface area-to-volume ratio. The relatively large volume creates a demand for oxygen that the relatively small surface area is unable to supply. Therefore, large organisms need specialised permeable surfaces whereby the absorption or exchange area is increased to satisfy the needs of the organism.

The rate at which an organism requires substances depends on its metabolic rate. An organism with a high metabolic rate has a high oxygen requirement and, therefore, possesses specialised, large gas exchange surfaces. This can be illustrated by comparing the mouse and the frog, which are organisms of similar size. The mouse has a metabolic rate and oxygen consumption approximately ten times greater than that of a frog, and has a proportionately bigger gas exchange surface. The frog's lungs are simple sacs, while the mouse has spongy lungs consisting of millions of microscopic alveoli. The large surface area is essential for the high rate of oxygen uptake required for the high metabolic rate of the mouse.

To summarise, an organism requires a specialised absorptive surface if it is terrestrial (with an impermeable surface), large (with a small surface area-to-volume ratio) or has a high metabolic rate.

Methods of increasing the area of an absorptive surface include:
- evagination (outfolding) of the surface
- invagination (infolding) of the surface
- flattening of the organism

Organisms with a flattened shape have a large surface area-to-volume ratio. A cube of $1 \times 1 \times 1$ arbitrary units has a surface area-to-volume ratio of 6. If this cube is flattened to dimensions of $0.1 \times 10 \times 1$ arbitrary units, it has the same volume but

Knowledge check 2

Calculate the surface area-to-volume ratio of a cuboid body of dimension $1 \times 2 \times 3$.

Gas exchange The process by which oxygen reaches cells and carbon dioxide is removed from them. It should not be confused with **respiration**.

Respiration The release of energy from food to synthesise ATP in all living cells.

Exam tip

Never say that small organisms have a large surface area — the surface area of a small organism is small! It is the surface area-to-volume ratio that is large.

Knowledge check 3

Give three reasons why humans need a specialised gas exchange surface (i.e. lungs).

the surface area-to-volume ratio is 22.2, an increase of nearly four-fold. Flattening not only increases the surface area-to-volume ratio, but it also decreases the distance over which substances have to be moved.

Skills development

Numeracy skills: calculations

To gain a better understanding, you should go over the calculation of surface area-to-volume ratios shown in Figure 1 and described above. Remember that the examination will require you to carry out at least one calculation, so practice is important. **Ratio** is calculated as the proportion of one value to another. Apart from ratio, you might also be asked to calculate magnification (see the first student guide in this series, covering AS Unit 1), percentage, percentage change and rate.

Percentage means out of 100. It is calculated by dividing a value by the total and multiplying by 100.

To calculate a **percentage change**, divide the difference (final value subtracted from the initial value) by the initial amount and multiply by 100. You might calculate a negative change — this represents a decrease.

The **rate** of a process is calculated as the change per unit time.

Exam tip

You have already met an absorptive surface — the one responsible for the absorption of the products of digestion by the villi (with microvilli further increasing surface area) in the ileum.

Table 1 shows some important examples of efficient absorptive surfaces in flowering plants and mammals.

Table 1 Absorptive surfaces in flowering plants and mammals

Absorptive surface	Structure	Function
Leaf mesophyll Palisade mesophyll, Spongy mesophyll, Air space system	The leaf is a flattened structure (its thinness ensures a short diffusion distance) with a tightly packed upper palisade mesophyll layer and a loosely packed lower spongy mesophyll layer	The wide expanse of palisade tissue is efficient at trapping light; the loose arrangement of the spongy layer provides an air space system through the leaf and creates a huge surface for gas exchange
Root hairs Epidermis, Root hair	Tubular extensions of the epidermal cells of the young root	Increase greatly the surface area of the root for the uptake of oxygen, water and ions

Absorptive surface	Structure	Function
Alveoli	Small (diameter 0.2 mm) sacs, occurring in clusters and in vast numbers within the mammalian lung; in human lungs there are 700 million, providing a total surface area of 70 m²	The huge, moist surface area provides for efficient gas exchange; alveolar walls are thin (0.1–1.0 µm), so the diffusion distance is short
Capillaries	Small (average diameter 8 µm), thin-walled blood vessels, with a total length of 100 000 km and surface area of 1000 m² in the human body	Extensive networks throughout the body represent a huge surface area for exchange of molecules between blood and body tissues; the number and distribution of capillaries are such that no cell is further away than 50 µm from a capillary
Red blood cells	Small (diameter around 8 µm), flexible biconcave discs, flattened and depressed in the centre, with a dumbbell-shaped cross-section	The biconcave disc shape greatly increases the surface area-to-volume ratio for efficient uptake of oxygen; the thinness of the cell, particularly where it is depressed in the centre, allows oxygen to diffuse to all the haemoglobin packed into the cell

Exam tip

Note that there are two gas exchange surfaces in mammals: between air in the alveoli and blood in the pulmonary capillaries within lungs; between blood in the systemic capillaries and cells in body tissues.

Knowledge check 4

Which two features of red blood cells provide them with a large surface area-to-volume ratio?

Exam tip

Questions often begin by asking for the definition of a term. As a part of your revision, construct a list of definitions for all the terms shown in bold in this and subsequent chapters.

The transport of substances

The molecules in gases and liquids move constantly and at random. If there is a difference in the concentration of molecules within an area, a net movement of the molecules occurs, resulting in the molecules becoming evenly distributed. This is illustrated in Figure 2.

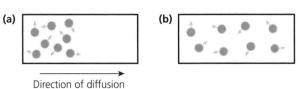

Direction of diffusion

Figure 2 Random movement of molecules causes them to become dispersed from an area of high concentration (a) until they are evenly distributed (b)

Such movements are known as **diffusion**. Diffusion can take place across surfaces as long as these are moist and permeable to the substances.

Diffusion is a sufficiently effective mechanism for the movement of substances across thin surfaces and within thin organisms. For example, a flatworm (an animal with a simple body plan — small, tapering and flattened — about 2 cm long and 2 mm thick) absorbs oxygen at its moist body surface and this oxygen diffuses to all body tissues since these are within easy reach of the surface. However, diffusion alone does not suffice for the movement of substances within large organisms.

Transport of substances in larger organisms occurs by **mass flow** (Figure 3). In mass flow, unlike diffusion, all molecules are swept along in the same direction.

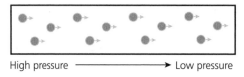

High pressure ⟶ Low pressure

Figure 3 The mass flow of molecules

Mass flow is brought about by a pressure difference. Mass flow systems include the xylem and phloem systems of flowering plants and the breathing (ventilation) and blood circulatory systems of mammals. Different mass flow systems have different means of generating a pressure difference (Table 2).

Table 2 Examples of mass flow systems in flowering plants and mammals

Mass flow system	Method of generating a pressure difference	Function of the mass flow system
Xylem system	Tension (negative pressure) in the leaf xylem generated by the transpirational loss of water from leaves	One-way transport of water and ions from roots to leaves in a flowering plant
Phloem system	Movement is driven by energy from the plant	Two-way flow of organic solutes (e.g. sucrose) in a flowering plant
Breathing (ventilation) system	Pressures in the thorax are alternately decreased (inducing inhalation) and increased (inducing exhalation)	Ventilation of the mammalian lungs, whereby air is alternately drawn in and forced out
Blood circulatory system	High pressure is generated by the muscular heart	Circulation of blood carrying oxygen, glucose, amino acids, fats, carbon dioxide, urea etc. in a mammal

There is an overlap between the principles of exchange and of transport. The flatness of an organism (or of a cell) increases its surface area-to-volume ratio and, therefore, its effectiveness in absorption. However, it also decreases the diffusion path, which is probably more important. An earthworm, though bigger than a flatworm, has a sufficiently large surface area-to-volume ratio to satisfy its gas exchange requirements, but needs an internal transport (blood circulatory) system, since the diffusion distance is too great. A flatworm has no need for an internal transport system, not only because it is thin, but also because its digestive system branches throughout its body so that food is absorbed in close proximity to all tissues (Figure 4).

Knowledge check 5

How are gases moved across exchange surfaces?

Knowledge check 6

List the features of a good transport system.

Knowledge check 7

Explain how thinness helps gaseous exchange *and* means that a transport system is not necessary.

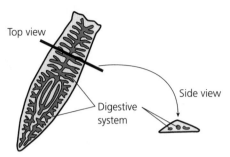

Figure 4 The flatworm: an animal with no need for a blood circulatory system

Exam tip

Note that reference has been made to flatworms, earthworms and even frogs, none of which is specified in the content of the unit. This is because you are required not only to have knowledge of the principles of exchange and transport, but to be able to apply your understanding in unfamiliar situations. Therefore, it is better to learn these principles through appropriate examples.

Summary

- An organism needs to absorb and/or exchange substances with the environment and these substances have to be distributed within the organism.
- Absorption or exchange of substances occurs at moist, permeable surfaces.
- A small organism has a large surface area-to-volume ratio, so can supply sufficient oxygen via its surface to satisfy the metabolic demands of its body (volume).
- An organism requires a specialised absorptive/ exchange surface if it is terrestrial (with an impermeable surface), if it is large (with a small surface area-to-volume ratio) or if it has a high metabolic rate.
- Specialised absorptive surfaces often involve outfoldings or infoldings, which increase surface area without increasing volume.
- Thinness increases the surface area-to-volume ratio and also reduces the diffusion path of substances.
- A transport system is required in larger animals and plants to distribute substances from one site to another.
- Transport systems involve mass flow, and a means of generating a pressure difference (often a pump).

Gas exchange

The surface over which gas exchange takes place must be:
- permeable to oxygen and carbon dioxide
- moist, since gases must dissolve in water before diffusing into tissue cells

Gas exchange across an exchange surface relies on diffusion. The rate of diffusion across this surface is influenced by a number of factors, as defined by **Fick's law**:

$$\text{rate of diffusion} \propto \frac{\text{surface area} \times \text{difference in concentration}}{\text{length of diffusion pathway}}$$

Fick's law shows that gas exchange is increased where:
- the exchange surface has a large surface area
- there is a big difference in the concentration of gases on each side of the surface
- the exchange surface is thin, with a short diffusion distance

The higher the metabolic rate of the organism, the greater the need for these principles to be met.

Gas exchange in flowering plants

There are two processes in flowering plants that involve gas exchange: **respiration** and **photosynthesis**. Respiration takes place in all tissues, all the time. Photosynthesis takes place only in green tissues (i.e. those containing chlorophyll), and only during the daylight hours. Indeed, the rate of photosynthesis depends on the light intensity. So, the maximum rate of photosynthesis occurs when the light intensity is highest, such as at midday. At this point, the rate of photosynthesis greatly exceeds the rate of respiration and so there is a net production of oxygen. At a specific low light intensity (during dawn and dusk) the rate of photosynthesis equals the rate of respiration and so the net exchange of oxygen is zero. This is known as the **compensation point**, since the rate of oxygen production (in photosynthesis) is balanced by the rate of oxygen consumption (in respiration).

These changes in oxygen use and release by a flowering plant are shown in Figure 5.

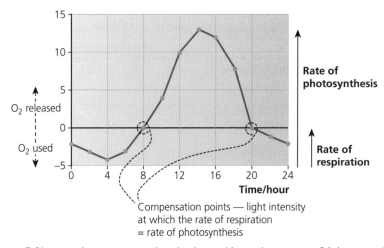

Figure 5 Changes in oxygen used and released by a plant over a 24-hour period

> **Exam tip**
>
> In Fick's law, the barrier to diffusion is often referred to as a membrane. The term membrane includes both the cell-surface membrane and the epithelium membrane, i.e. the layer of cells covering an organ.

Respiratory gas exchange in flowering plants

The roots of plants use energy in processes such as cell division (growth) and the active transport of ions from the soil solution. In the growth region, the epidermal cells possess **root hairs**, which increase the surface area-to-volume ratio. In soil that is not waterlogged, root hairs are surrounded by air spaces between the particles of soil. Diffusion of respiratory gases occurs through the cell wall and cell membrane of these root hairs (Figure 6).

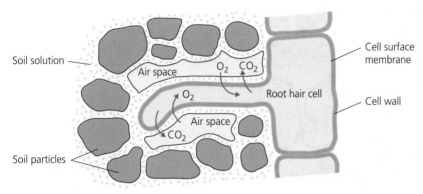

Figure 6 Gas exchange in a root hair cell

In stems, especially those of woody plants, the most active cells are those under the surface. Although the stem's outer covering is waterproofed to reduce water loss to the air (and so is impermeable to gases) there are small pores to allow oxygen in and carbon dioxide out.

Plants lack specialised respiratory surfaces and yet can be very large. This is because they lack tissues with a high energy demand and so have low respiration rates.

Photosynthetic gas exchange in flowering plants

The **leaf** is the major photosynthetic organ in a flowering plant. The leaf needs a specialised gas exchange surface because:

- a high rate of photosynthesis is generated
- the concentration of carbon dioxide in air is low (400 parts per million or 0.04%)

The leaf epidermis, particularly the lower epidermis, possesses **guard cells** that control the opening and closure of **stomata**. The stomata open during the hours of daylight. When open, air, containing carbon dioxide, diffuses into and out of the leaf mesophyll. Carbon dioxide diffuses through an **air space system** provided by the **spongy mesophyll**. Having diffused through the air space system, carbon dioxide is absorbed by mesophyll cells (in which carbon dioxide concentration is low as it is used in photosynthesis). It is this moist mesophyll surface that represents the gas exchange surface (Figure 7). Since the leaf is *broad* and *thin*, there is a large surface area and the diffusion distance for gases is short.

Exam tip

Don't forget that plants need oxygen for respiration. This is particularly crucial for roots and is the reason that plant growth and survival is so dependent on the soil containing air and thus having a good crumb structure.

Knowledge check 8

Suggest why plants have a lower metabolic rate than animals.

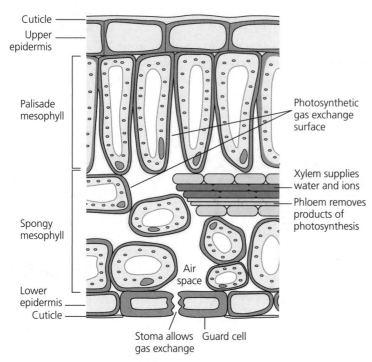

Figure 7 The structure of a leaf

Oxygen produced in photosynthesis diffuses out of mesophyll cells into the air space and then out through open stomata. Some oxygen is, of course, used up in respiration.

Plants living in water: hydrophytes

There is much less oxygen dissolved in water than there is oxygen in air, so the stems and leaves of aquatic flowering plants (hydrophytes) have adaptations to facilitate the uptake and movement of oxygen and carbon dioxide (Figure 8).

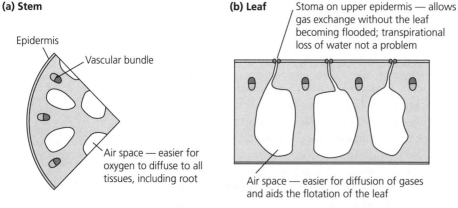

Figure 8 Adaptations of a hydrophyte (a) stem and (b) leaf

Gas exchange in mammals

Gas exchange in cells occurs by diffusion (Figure 9). In cells respiring aerobically, oxygen is used and carbon dioxide is produced. This affects the concentration gradients and so oxygen diffuses into the cell and carbon dioxide diffuses out.

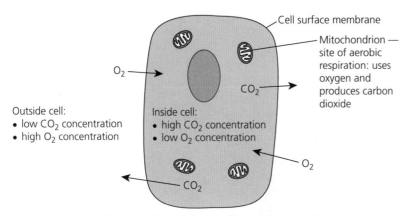

Figure 9 Gas exchange in a respiring cell

The movement of oxygen to cells, and of carbon dioxide out, involves four stages:

- diffusion of gases between respiring cells and blood
- transport of gases in blood
- diffusion of gases across the gas exchange surface between alveolar air and blood
- ventilation of the lungs with fresh air

The structure of the lungs

Air is breathed through the nostrils or mouth, and enters or leaves the lungs via the **trachea**. The lungs are situated within the **thorax** (also known as the **thoracic cavity**). The trachea branches into two **bronchi** (singular **bronchus**) which further branch into a series of ever-finer **bronchioles** forming a **bronchial tree**. Each **terminal bronchiole** leads to a cluster of **alveoli**, with an **alveolar duct** connecting with each **alveolus**. Each individual alveolus is tightly wrapped in blood capillaries and it is here that gas exchange takes place. The structure of the lung system is shown in Figure 10.

> **Exam tip**
>
> Note that, in mammals, there are two mass flow systems involved in the movement of oxygen, though movement at the two interfaces relies on diffusion.

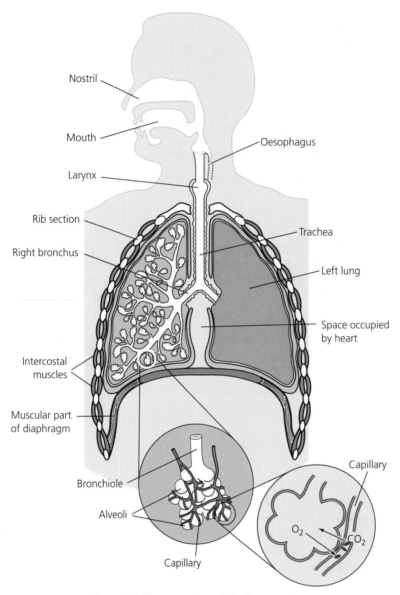

Figure 10 The structure of the lung system

Ventilation of the lungs

Changing the volume of the thoracic cavity changes the air pressure inside the lungs. Air moves from a region of high pressure to a region of low pressure. When the volume of the thorax is increased, the pressure in the lungs is decreased, becoming lower than atmospheric pressure, so that air moves into the lungs (inhalation or inspiration). When the volume of the thorax is decreased, the pressure in the lungs is increased, becoming higher than atmospheric pressure so that air moves out of the lungs (exhalation or expiration). The mechanisms of **inhalation** and **exhalation**, during normal breathing, are summarised in Figure 11.

Exam tip

Be careful about the use of terms. **Respiration** occurs in cells and is the release of energy from food to make ATP. **Breathing** refers to the movements of the ribs and diaphragm that cause air to enter and leave (ventilate) the lungs.

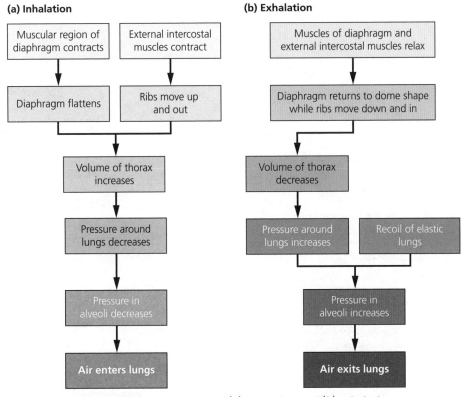

(a) Inhalation

(b) Exhalation

Figure 11 The mechanism of (a) inhalation and (b) exhalation

During forced exhalation, such as during hard exercise and coughing, the volume of the thoracic cavity is further reduced by the contraction of the internal intercostal muscles depressing the rib cage.

Alveolar structure and gas exchange

In each human lung there are about 350 million alveoli, each of which is about 200 μm in diameter. Their total surface area is about 70 m². Each alveolus is lined with a single layer of **squamous epithelial cells** only 0.05–0.3 μm thick. The inner surface of each alveolus is moist. Around each alveolus is a network of blood capillaries. These capillaries also have walls consisting of a squamous (also called pavement) epithelium (0.04–0.2 μm thick).

Gases can diffuse across the moist, permeable walls of the alveoli. Diffusion of gases between the alveoli and the blood is rapid because:

- concentration gradients are maintained by ventilation of the lungs and blood flow through pulmonary capillaries
- the alveoli and pulmonary capillaries have very large surface areas
- the walls of both alveoli and capillaries are thin and therefore the diffusion distance is very short — the distance between alveolar air and the red blood cells averages only about 0.5 μm

Knowledge check 11

Explain the advantage of ventilating the lungs.

Exam tip

The alveoli of the lungs exhibit all the features of an efficient gas exchange surface: permeable, moist, large surface area, thin walls, close to transport systems.

Knowledge check 12

Explain the statement 'inhalation is an active process but exhalation is usually passive'.

The structure of an alveolus and associated capillaries is shown in Figure 12.

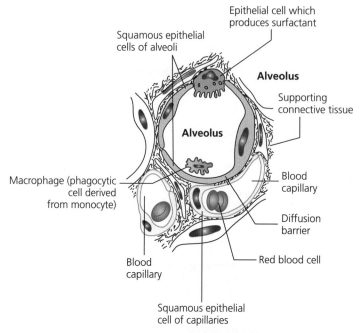

Figure 12 A section through several alveoli and capillaries

Since the alveolar surface is situated deep inside the body, evaporation of water from its moist surface is reduced to a minimum. Other cell types are present in the alveolar wall. **Surfactant-secreting cells** (septal cells) produce a detergent-like substance that reduces surface tension in the fluid coating the alveoli, and without which the alveoli would collapse due to cohesive forces between the water molecules lining the air sacs. **Macrophages** (derived from **monocytes**, a type of white blood cell) protect the lungs from a broad spectrum of microbes and particles by ingesting them through phagocytosis. Elastic fibres are also associated with the alveolar walls, and the elastic recoil of the alveoli helps to force air out during exhalation.

Smoking and lung disease

Tobacco smoke contains thousands of toxic substances, many of which are collectively known as tar. As a result, smoking has been linked to a number of lung diseases. Smokers frequently develop a combination of **chronic bronchitis** and **emphysema**, known as chronic obstructive pulmonary disease (COPD).

Tar brings about an inflammatory response in which airways narrow and excessive amounts of mucus are produced. Furthermore, tar paralyses the cilia that sweep mucus and bacteria away from the lungs, so pathogens and mucus build up. This leads to phlegm production, coughing and breathlessness — symptoms of chronic bronchitis. Inability to clear mucus and bacteria results in an increased susceptibility to chest infections, including pneumonia.

Exam tip

Alveoli and capillaries both have thin walls. However, it is not sufficient just to say that the walls are one cell thick. You must also say that they consist of squamous (pavement) epithelium.

Knowledge check 13

Babies born prematurely are often deficient in surfactant, causing a condition called 'respiratory distress syndrome'. From what you know about the role of surfactant, describe and explain the symptoms of this syndrome.

In emphysema, the inflammatory response to smoke inhalation leads to breakdown of alveolar walls. This reduces the area available for gas exchange so it becomes difficult to get enough oxygen. There is also a loss of elastic fibres in the alveolar walls. Therefore, exhalation becomes more difficult because the ability of alveoli to recoil following inhalation is reduced.

Tobacco smoke contains many **carcinogens** — substances that can induce **cancer** — of which tar is the most important. Carcinogens can damage DNA in the cells lining the bronchial tubes. Cells with DNA damage may divide in a modified and uncontrolled way, producing a mass of unspecialised cells known as a tumour. A cancerous or malignant tumour may spread to invade other tissues.

Knowledge check 14

Explain why a person with emphysema would have difficulty undertaking strenuous activity.

Practical work

Understand the use of a simple respirometer:
- measure oxygen consumption (with potassium hydroxide present)
- measure the net difference between carbon dioxide production and oxygen consumption (in the absence of potassium hydroxide) and so determine carbon dioxide production

Summary

- Gas exchange is most efficient when the exchange surface has a large area, a large diffusion gradient is maintained and the barrier to diffusion is thin.
- In plants, during the night (when only respiration can take place) oxygen is absorbed and carbon dioxide is released. During the day, when the rate of photosynthesis exceeds the rate of respiration, carbon dioxide is absorbed and oxygen is released.
- In plant leaves, the mesophyll is the main gas exchange surface. It has a large surface area, is thin, and has an extensive air-space system connected to the atmosphere by stomata that allow the entry of air to maintain a concentration difference.
- In mammals (humans), the alveoli in the lungs are the specialised gas exchange surface through which oxygen diffuses into blood and carbon dioxide diffuses out.
- The millions of sac-like alveoli provide a large surface area.

- The extremely thin walls of alveoli and capillaries (consisting only of squamous epithelium) provide a short diffusion distance.
- The ventilation of the lungs and the circulation of blood across the alveolar surface both maintain high concentration gradients between the alveolar air and blood.
- Ventilation occurs because breathing movements create pressure differences between the atmosphere and the air in the lungs. Inhalation involves the contraction of the external intercostal muscles and the diaphragm muscle, so increasing the volume of the thorax and decreasing pressure; exhalation involves the relaxation of these muscles and the recoil of elastic alveoli.
- A number of diseases are associated with smoking, including chronic obstructive pulmonary disease, which is caused by a combination of chronic bronchitis and emphysema, and lung cancer.

Transport and transpiration in plants

Vascular tissues

There are two transport systems in flowering plants:

- **xylem** for water and inorganic ions
- **phloem** for organic molecules, such as sucrose and amino acids

These systems are contained within the plant's **vascular tissues**. Vascular tissues form a central **stele** (vascular cylinder) in the root, but peripheral **vascular bundles** in the stem. In the leaf, vascular tissues form a central large vascular bundle (midrib) from which smaller vascular bundles (veins) run through the leaf (Figure 13).

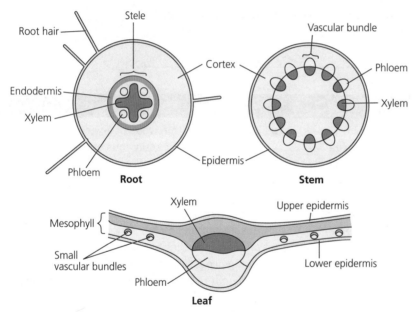

Figure 13 The distribution of vascular tissue (xylem and phloem) in cross-sections of root, stem and leaf

Knowledge check 15

Describe the distribution of:

a xylem

b phloem

in both roots and stems.

Xylem

Xylem contains different types of cell. Those that transport most of the water and ions are called **xylem vessels**. In vessels, a secondary wall, impregnated with **lignin**, is formed inside the primary cellulose wall. Lignin is impermeable to water, so mature xylem vessels are dead.

There are different patterns of lignification in xylem vessels (Figure 14). In the first formed xylem, known as the **protoxylem**, found in the growing regions behind the root and shoot tips, an **annular** or **spiral** pattern is produced. These patterns allow the vessels to elongate along with other tissues in the growth regions. In the xylem found in the mature parts of the plant, known as the **metaxylem**, there is a greater deposition of lignin and a **reticulated** or **pitted** pattern is produced. Reticulated vessels are thickened by interconnecting bars of lignin; pitted vessels are uniformly thickened, except at pores seen as pits that allow rapid movement of water and ions out of vessels to surrounding cells. The role of lignin is to prevent vessels from collapsing when under tension. As the vessels form, their end-walls break down, so continuous tubes are formed. As a result, water movement through xylem vessels requires less pressure than through living cells, where movement would be slowed down by cell contents.

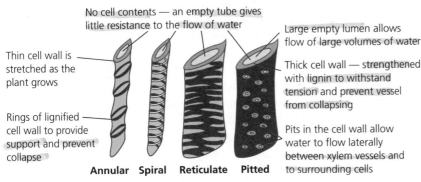

Figure 14 Different patterns of lignification in xylem vessels

Phloem

Phloem tissue consists of different cell types (Figure 15). The transporting cells are the **sieve tube elements**. These lie end-to-end to form a continuous stack — the sieve tube. The thin cellulose walls at the ends of the cells are perforated to form **sieve plates**, so making movement between sieve tube elements easier. **Companion cells** are closely associated with sieve tube elements. The cytoplasm of companion cells is linked via plasmodesmata to that of sieve tube elements. The smaller companion cells have a dense cytoplasm with a large number of mitochondria and have high levels of metabolic activity (p. 25). They maintain the activity of the sieve tube elements. This is important since sieve tube elements lose their nuclei and many organelles as they mature.

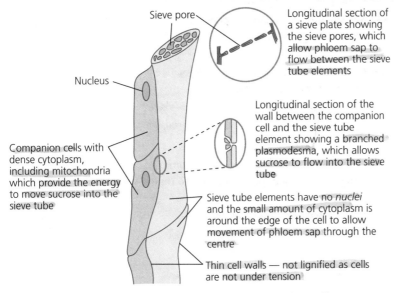

Figure 15 Phloem sieve tube and companion cells

Exam tip

You may be asked to explain why xylem vessels are lignified. Do not say that it strengthens xylem vessels; its primary role is to prevent xylem vessels collapsing under the negative pressures (tension) created as a result of transpiration. A secondary function of lignin is that it offers support to the plant.

Knowledge check 16

List four features of xylem that make it well adapted to the transport of water.

Knowledge check 17

Explain the role of companion cells in translocation.

Transpiration and the movement of water and ions

Transpiration is the process whereby water vapour is lost from plants. One of the main functions of the waxy cuticle is to waterproof the leaf's surface and so reduce evaporation of water from the epidermis. While being an efficient barrier, the cuticle is not impermeable to the passage of water and some water is lost by **cuticular transpiration**. However, water evaporates readily from the cell walls of mesophyll cells, and so water vapour accumulates in the air spaces. Stomata are closed at night, so little of this water vapour escapes. During the day, stomata are open to allow the inward diffusion of carbon dioxide for photosynthesis. As a consequence, water vapour diffuses out. This **stomatal transpiration** is the combined effect of evaporation from the mesophyll surface and diffusion of water vapour out of open stomata. Stomatal transpiration accounts for 90% or more of all water loss in most plant species.

Internal factors that affect the rate of transpiration are:

- leaf surface area — increased surface area increases the rate of transpiration because more surface with stomata is exposed to the air
- stomatal density — the greater the density, the more stomata (per unit area) there are for water to diffuse out of the leaf
- cuticle thickness — the thicker the cuticle, the more effective it is at waterproofing and, therefore, reducing cuticular transpiration

A number of **external factors** influence transpiration rate:

- **Temperature** High temperatures increase the transpiration rate in two ways: greater heat energy increases evaporation of water from the walls of mesophyll cells and increases diffusion of water molecules out of open stomata.
- **Air movements** Increased air movement outside the leaf increases transpiration rate by blowing away the diffusion shells of water vapour that gather outside open stomata. This increases the water potential gradient out of the leaf.
- **Humidity** Increased humidity decreases the rate of transpiration because when the atmosphere holds more water molecules the water potential gradient to the outside is reduced.
- **Light** Transpiration occurs more quickly in bright light than in the dark since stomata, the major route by which water vapour is lost, are open.
- **Soil water availability** If water is in short supply, plants are unable to replace water lost in transpiration with water from the soil, so the stomata close, reducing transpirational loss.

As water evaporates from the mesophyll cell walls, it is replaced by water from the xylem vessels in the leaf. Flow of water through the mesophyll occurs by two main pathways:

- **apoplast pathway** — water moves along the cellulose fibrils of cell walls
- **symplast pathway** — water diffuses from cell cytoplasm to cell cytoplasm through plasmodesmata

Most of the water follows the apoplast pathway since this offers least resistance.

> **Exam tip**
>
> It is incorrect to say that the cuticle prevents water loss — it *reduces* water loss. Xerophytes have a *thick* waxy cuticle to *further* reduce water loss.

> **Exam tip**
>
> It is not accurate to say that water evaporates out of the stomata — it evaporates from the mesophyll surface and diffuses out of the stomata.

As water is drawn out of xylem vessels, a **tension** (negative pressure) develops and a **water potential gradient** is established through the plant. Water moves through the plant because of this water potential gradient and because of **cohesive forces** between water molecules — the water column in the xylem vessel moves 'as one'. Further, there are **adhesive forces** between water and xylem vessel walls, which tend to prevent the water column dropping down under gravity. This mechanism of water movement through the plant is described by the **cohesion–tension theory**. The pull caused by transpiration is great enough to cause the circumference of a tree to become smaller when transpiration is at its maximum (i.e. midday).

Water and ions are taken up at the root epidermis, where root hair cells greatly increase the surface area. There are two routes:

- Water is absorbed into root hair cells by osmosis; ions are absorbed by active transport (or, if the diffusion gradient is favourable, by facilitated diffusion). Subsequent movement across the cortex is along the symplast pathway.
- Water and ions are adsorbed onto the cellulose fibrils of cell walls. Subsequent movement across the cortex is along the apoplast pathway.

Transport of water across the **cortex** of the root occurs in the same way as transport through the mesophyll of the leaf, i.e. mostly by the apoplast pathway and some by the symplast pathway. However, the cells of the **endodermis** have a **Casparian strip** that completely encircles each cell and is impermeable to water. This is a barrier to the apoplast pathway, so water can *only travel via the symplast pathway into xylem vessels*. This allows active control of the passage of water and dissolved ions. Ions are pumped by the cytoplasm of the endodermal cells into the xylem and water follows along the resultant water potential gradient, thus creating **root pressure**. Water also moves into xylem due to the transpirational pull.

> **Knowledge check 18**
>
> Explain why a lack of oxygen reduces ion uptake.

> **Knowledge check 19**
>
> Explain why a respiratory poison would reduce the effect of the root pressure.

> **Exam tip**
>
> Always refer to water diffusing down a water potential gradient — *never* use the phrase 'water concentration gradient'. Water diffuses through the root cortex, moves up the xylem and diffuses as water vapour out of the stomata — in each case from a region of higher water potential (less negative) to a region of lower water potential (more negative).

> **Exam tip**
>
> Transpiration represents a loss of water to the plant. This is only a problem if the loss is excessive or if soil water becomes unavailable, as in a drought. Transpiration is the major force drawing water (containing dissolved ions) up the plant and it therefore results in a continuous supply of ions for use in the leaf.

The movement of water out of a leaf, across a root and ultimately through the whole plant is shown in Figure 16, along with the water potential gradient from soil (less negative ψ) to air (more negative ψ).

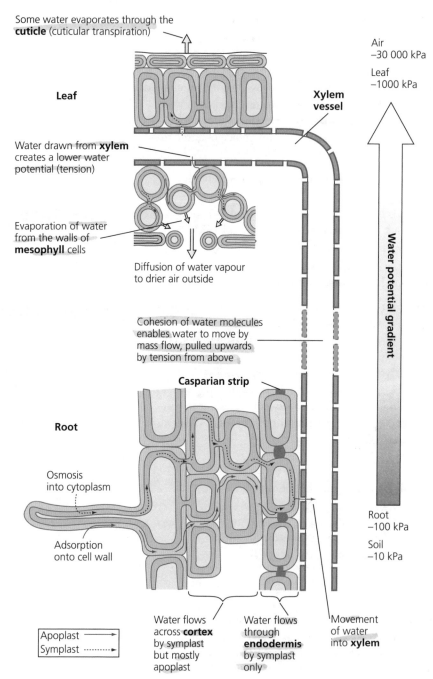

Air
−30 000 kPa

Leaf
−1000 kPa

Some water evaporates through the **cuticle** (cuticular transpiration)

Leaf

Xylem vessel

Water drawn from **xylem** creates a lower water potential (tension)

Evaporation of water from the walls of **mesophyll** cells

Diffusion of water vapour to drier air outside

Water potential gradient

Cohesion of water molecules enables water to move by mass flow, pulled upwards by tension from above

Casparian strip

Root

Osmosis into cytoplasm

Adsorption onto cell wall

Root
−100 kPa

Soil
−10 kPa

Apoplast ──────▶
Symplast ┄┄┄┄▶

Water flows across **cortex** by symplast but mostly apoplast

Water flows through **endodermis** by symplast only

Movement of water into **xylem**

Figure 16 The movement of water through a plant

Xerophytes

Xerophytes are plants that are adapted to living in dry habitats. Some of the features of xerophytic plants are summarised in Table 3.

Table 3 Summary of some xerophytic features

Feature	Advantage of feature
Thick cuticle	Increases efficiency of waterproofing layer
Few stomata, leaf very small or with a reduced surface area-to-volume ratio	Reduced pathway for water loss
Stomata sunken into leaf; leaf covered by hairs; leaf rolled with stomata on the inside	Humid air builds up immediately outside stomata reducing diffusion gradient out of the leaf
Water storage cells in leaf or stem — succulence	Greater storage of water for use during periods of drought
Leaves adapted as spines	Spines prevent grazing and exposure of plant tissue to evaporation
Extensive shallow root network or deep network of roots	Greater uptake of water when it does become available

Knowledge check 20

With reference to the water potential gradient, explain how sunken stomata reduce transpiration.

Marram grass is well adapted to live in sand dunes — a dry and windy environment. A cross-section of its leaf is shown in Figure 17.

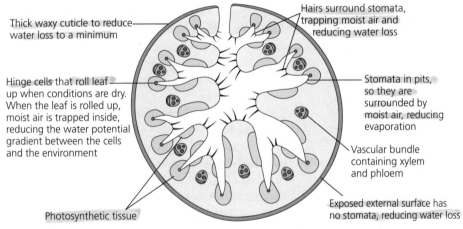

Figure 17 Marram grass is well adapted to restrict water loss

Translocation of organic solutes

Movement of organic solutes within the phloem sap is known as **translocation**. Phloem sap contains primarily **sucrose** (the transport carbohydrate in plants), though amino acids and other solutes are also present. There are two main principles with respect to movement of phloem sap:

- Movement involves **energy expenditure**. There are several lines of evidence that mass flow is maintained by an active mechanism:
  - The rate of flow ($1 \, \text{m} \, \text{h}^{-1}$) is higher than can be accounted for by diffusion.
  - Companion cells have a particularly high density of mitochondria and are more metabolically active than other plant cells. ATP is used to pump sucrose into companion cells from where it enters the sieve tube element via plasmodesmata.
  - Metabolic poisons (e.g. potassium cyanide), which stop respiration, also stop translocation.

Exam tip

The transport of water is a passive process, whereas the transport of organic solutes is an active process.

- Movement is **two-way** or, more precisely, movement occurs from '**source to sink**'. The source is the organ where sugar is produced in photosynthesis or by the breakdown of starch (e.g. leaves) and the sink is the organ that consumes or stores carbohydrate (developing buds, flowers, fruit, roots and root storage organs). So sugar can be moved up to a developing bud at the shoot tip or down to the roots. Furthermore, source and sink depend on the season: a potato tuber is a sink as it builds up stores of carbohydrate in the summer but is a source in the spring when starch is broken down to supply the energy for the growth of shoots. Evidence for two-way flow comes from the use of radioactively labelled sucrose: following the supply of labelled sucrose to a mature leaf, radioactivity is detected in the shoot tip above and in the roots below.

Knowledge check 21

Experiments with labelled sucrose show that radioactivity is detected in very young leaves but not in mature leaves. Explain why.

Exam tip

The topic lends itself to the interpretation and/or drawing of a photograph. You should examine photographs and/or microscope slides of sections through a root and stele, stem and vascular bundle, and leaves, especially xerophytic leaves.

Exam tip

You are not required to give details of any of the theories of translocation.

Summary

- Flowering plants have two transport (vascular) systems: xylem and phloem.
- In roots, xylem and phloem occur in a central vascular cylinder or stele; in stems, in radially arranged vascular bundles that branch into the leaves.
- In xylem, water and ions are transported in vessels with appropriate adaptations — no cross walls or cytoplasm (to impede movement), lignified (to prevent collapse) but with pores (to allow water to move out laterally).
- Water and ions are absorbed by root hairs (with large surface area) and are passed across the root cortex by apoplast and symplast pathways.
- At the endodermis that lines the stele, water and ions are forced to enter the cytoplasm (symplast pathway). From here ions are pumped into the xylem so that water follows osmotically, creating root pressure.
- Xylem carries water and ions from the roots through the stem to the leaves — it is one-way flow.
- Water is lost from leaves by transpiration, whereby water evaporates from the mesophyll surface and diffuses out, mainly through open stomata.

- Transpiration is influenced by internal factors (e.g. stomatal density and cuticle thickness) and external factors (e.g. temperature, air movement, humidity and light).
- Xerophytes have adaptations that reduce transpiration.
- Transpiration creates a negative pressure (tension) in the leaf xylem and is the major force drawing water up, supported by the combined forces of cohesion and adhesion, and to a lesser extent by the root pressure.
- In all cases water moves down a water potential gradient, i.e. from a region of higher water potential to a region of lower (more negative) water potential.
- In phloem, organic solutes (mainly sucrose and amino acids) are transported in sieve tubes. These have reduced cytoplasm and cross walls with pores, the sieve plates, to facilitate movement. Translocation of organic solutes is an active process that relies on the metabolism of adjacent companion cells.
- Phloem carries organic solutes from the 'source' of sugar (mature leaves) to the 'sinks' (roots and growth areas).

■ Circulatory system in mammals

The double circulatory system

In mammals, during one complete pathway around the body, blood flows through the heart twice. One circuit going from the heart to the lungs and back is the **pulmonary circulation**. The other circuit from the heart to the rest of the body is the **systemic circulation**. Since there are two circuits, this is called a **double circulatory system**. Blood is pumped twice by the heart, from the *right ventricle to the lungs* and from the *left ventricle to the body*. These two chambers both have thick walls made of cardiac muscle, but the muscle of the left ventricle is thicker than the muscle of the right ventricle. Therefore, the blood pumped by the left ventricle is pushed harder and is under greater pressure than blood from the right ventricle. This two-pressure system has advantages:

- The low pressure in the pulmonary circulation pushes blood slowly to the lungs allowing more time for gas exchange.
- The high pressure in the systemic circulation ensures blood is pumped to all the other body organs and allows tissue fluid to form in each organ.

An artery branches off the systemic circulation to supply each of the body organs, and a vein returns the blood to the heart. The main blood vessels of the thorax and abdomen are shown in Figure 18.

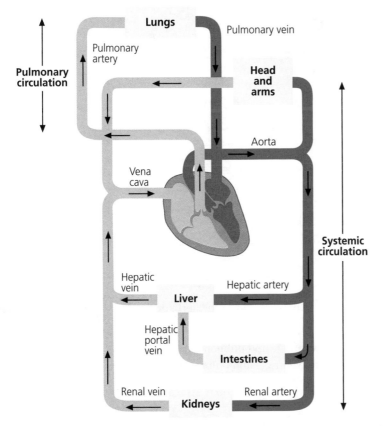

Figure 18 The double circulatory system, showing the main blood vessels

Knowledge check 22

Explain why large, active animals need a transport system.

....

Knowledge check 23

What are the advantages of a double circulatory system over a single circulatory system?

....

Exam tip

You are required to name only the main blood vessels, as shown in Figure 18.

....

Knowledge check 24

Describe the pathway that blood takes between the vena cava and the aorta.

....

The heart needs its own supply of blood — the coronary circulation — to provide cardiac muscle with oxygen and nutrients. The coronary arteries (Figure 19) arise from the base of the aorta. Problems may occur in branches of the coronary arteries (pp. 37–39).

The heart

The structure of the heart

The structure of the heart with its associated blood vessels is shown in Figure 20.

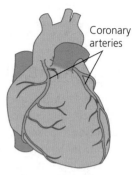

Figure 19 Coronary arteries supply blood to the heart muscle

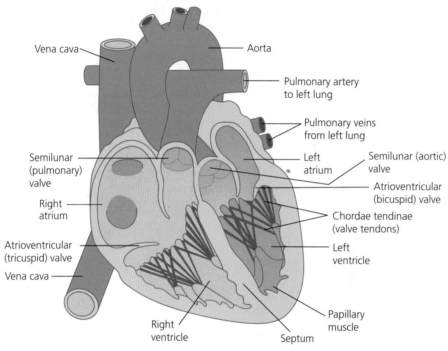

Figure 20 The structure of the heart

The cardiac cycle

There are three main stages during each beat of the heart:

- **atrial systole** — atria contract (ventricles are relaxed)
- **ventricular systole** — ventricles contract (atria are relaxed)
- **diastole** — both atria and ventricles are relaxed

These stages occur at the same time in the right and left sides of the heart.

The **pressure changes** in the left atrium, the left ventricle and in the aorta during one cardiac cycle are shown in Figure 21.

A graph of the changes in the right atrium, right ventricle and pulmonary artery shows all the same features, but the pressures in the right ventricle and pulmonary artery are lower.

Knowledge check 25

Why is the maximum pressure similar in both atria?

As valves close, the flaps of tissue snap together, making a sound. The first (and softer) heart sound is produced by closure of the atrioventricular valves; the second, sharper sound is produced by the clapping shut of the semilunar valves. These sounds are shown in the **phonocardiogram** in Figure 21. The electrical activity through the heart is recorded as an **electrocardiogram**, also shown in Figure 21. Wave **P** shows the excitation of the atria, **QRS** indicates the excitation of the ventricles and **T** corresponds to diastole.

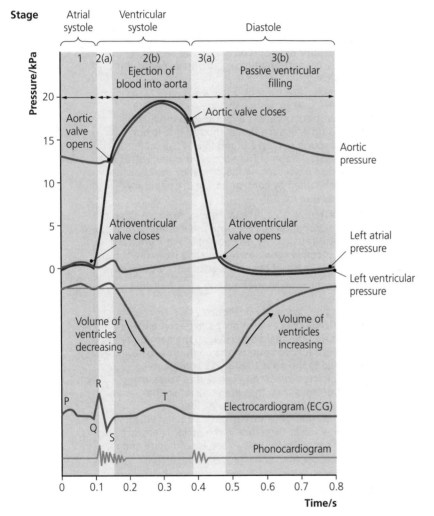

Figure 21 The pressure changes in left atrium, left ventricle and aorta during one cardiac cycle (with ventricular volume changes, electrocardiogram and phonocardiogram)

The sequence of events during each phase of the cardiac cycle is shown in Table 4. The numbers of the stages in Table 4 are also shown in Figure 21 above.

Exam tip

You should be able to interpret data showing pressure changes in the heart chambers and major arteries, such as those in Figure 21. You need to work through this graph and recognise the events of the cardiac cycle. Note that blood flows from a region of high pressure to a region of lower pressure, unless prevented from doing so by the forced closure of a valve. When valve flaps snap together a heart sound is made.

Knowledge check 26

What causes the atrioventricular valves to close?

Knowledge check 27

During which stage(s) in the cardiac cycle are the semilunar valves closed? Explain your answer.

Knowledge check 28

Figure 21 shows a heartbeat of duration 0.8 seconds. Calculate the heart rate in beats per minute.

Table 4 The sequence of events during stages of the cardiac cycle

Stage	Description of events
(1) Atrial systole	Atria contract (ventricles are relaxed) pushing more blood into the ventricles. This is essentially topping up the ventricles (since blood has already entered the ventricles during diastole, when the atrioventricular valves are open).
(2) Ventricular systole (a) (b)	Ventricles contract (atria are relaxed) causing the pressure of the blood inside the ventricles to become greater and forcing the atrioventricular valves shut. Two phases are evident (Figure 21): (a) Ventricular pressure increase causes the atrioventricular valves to bulge into the atria increasing pressure there, though this is not great enough to cause blood to exit the major arteries; the flaps of the atrioventricular valves are prevented from turning inside out by the chordae tendinae, aided by contraction of the papillary muscles (Figure 20) (b) Ventricular pressure increases to exceed that in the major arteries, pushing the semilunar valves open and causing the ejection of blood from the heart; blood is returned to the atria from the major veins and so atrial pressure gradually increases
(3) Diastole (a) (b)	Cardiac muscle throughout the atria and ventricles relaxes. Two phases are evident (Figure 21): (a) Ventricular pressure drops to become lower than that in the main arteries, so the semilunar valves are forced shut. Blood continues to be returned to the atria though it cannot enter the ventricles since the ventricular pressure is still greater than that in the atria, so the atrioventricular valves remain closed. (b) Ventricular pressure drops to a point where it becomes lower than that in the atria. Therefore, the atrioventricular valves are forced open and blood enters the ventricles from the atria.

Coordination of the cardiac cycle

The sequence of atrial systole, ventricular systole and diastole occurs as a result of coordinated waves of excitation through the heart. The heartbeat is **myogenic** — contraction originates in the heart itself and does not depend on nervous stimulation. The heartbeat starts at the **sinoatrial node** (SA node), a small patch of tissue in the right atrium that acts as a **pacemaker**. Excitation from the SA node spreads rapidly over the atria, causing them to contract together (atrial systole).

Between the atria and the ventricles is a layer of **non-conductive tissue**, which prevents the spread of the wave of excitation passing directly from the atria to the ventricles. The

Knowledge check 29

List the functions of the SAN, AVN and the bundle of His.

only conducting route for the wave of excitation to the ventricles is via the **atrioventricular node** (AV node). Waves of excitation from the SA node reach the AV node and there is a short time delay before the waves of excitation pass down to the base of the ventricles.

The AV node is connected to specialised muscle fibres called the **bundle of His**, which transports a wave of excitation down both sides of the septum to the base of the ventricles. Special fibres called **Purkinje fibres** (also known as Purkyne fibres), which branch upwards, conduct the waves of excitation to all parts of the ventricles, causing them to contract from the bottom up. This ensures that blood is pushed up from the ventricles into the arteries. The two ventricles contract at exactly the same time (ventricular systole). Once cardiac muscle contracts it goes into a period of rest (diastole) before it can contract again. The waves of excitation through the heart are shown in Figure 22.

Knowledge check 30

Explain the advantage of:

a the delay in the impulses through the AV node

b the ventricles contracting from the base upwards

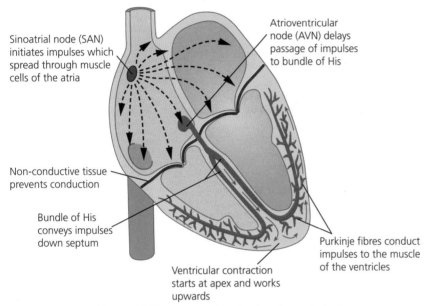

Figure 22 The waves of excitation through the heart

Labels:
- Sinoatrial node (SAN) initiates impulses which spread through muscle cells of the atria
- Atrioventricular node (AVN) delays passage of impulses to bundle of His
- Non-conductive tissue prevents conduction
- Bundle of His conveys impulses down septum
- Ventricular contraction starts at apex and works upwards
- Purkinje fibres conduct impulses to the muscle of the ventricles

The blood vessels

Arteries are adapted for carrying blood under high pressure away from the heart, towards individual organs.

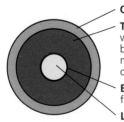

- **Outer layer of fibrous tissue** — protection
- **Thick middle layer containing elastic tissue** — allows stretching when blood surges (during systole) and recoils to continue pushing blood along (during diastole) — and **muscle tissue** — may contract to narrow the lumen (vasoconstriction) so reducing blood supply to an organ, or relax to increase supply (vasodilation)
- **Endothelium** — provides a smooth inner surface which reduces the friction caused by blood flow through the lumen
- **Lumen** — small, maintaining a high blood pressure, and may be constricted or dilated

Figure 23 The wall of an artery

Capillaries are adapted for the exchange of material between blood and tissue cells. The vast networks of capillaries slow the blood, giving time for diffusion to occur.

Exam tip

Do *not* say that the muscle in arterial walls is used to push the blood along. Its contraction *constricts* (narrows) the artery.

Knowledge check 31

Which possesses more elastic tissue, the aorta or the pulmonary artery? Explain your answer.

The fluid that leaks out of capillaries is called **tissue fluid**. It bathes all the surrounding cells. Polymorphs and monocytes, two types of white blood cell, can also leave capillaries, squeezing between adjacent endothelial cells, at sites of infection.

Squamous (pavement) endothelium — thin wall, permeable to water and solutes, so providing a short diffusion distance and facilitating the exchange of substances between the blood and tissue cells

Figure 24 A capillary

Veins are adapted for carrying blood under low pressure and returning it to the heart. They have semilunar valves to prevent backflow of blood. Blood is squeezed along when skeletal muscles contract.

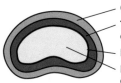

Outer layer of fibrous tissue — protection

Thin middle layer containing some smooth muscle and few elastic fibres — small layer since blood is under low pressure

Endothelium — provides a smooth inner surface

Lumen — large space making it easier for blood to enter from the capillaries while friction is reduced as blood flows back to the heart

Figure 25 The wall of a vein

Tissue fluid

Tissue fluid is formed at the arterial end of capillaries because the blood has a high **hydrostatic pressure** (due to pumping of the heart). This pressure forces plasma, minus its proteins, out of capillaries (proteins are too big to filter out) and is only partly resisted by the low solute potential of the blood (resulting from retention of proteins). Tissue fluid carries glucose, ions and oxygen to the cells. The loss of fluid from capillaries reduces the hydrostatic pressure of the blood. Therefore, at the venous end, due to the low solute potential of the blood, fluid (i.e. water, carbon dioxide and other waste products) is drawn into capillaries by osmosis and diffusion.

Blood

Blood is a suspension of cells in a pale yellow liquid called **plasma**.

Blood cells

Red blood cells are much more numerous than white blood cells, of which there are different types — **polymorphs**, **monocytes** and **lymphocytes** (Figure 26).

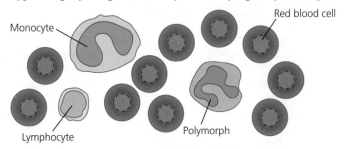

Figure 26 A drawing of the different cell types in a blood smear

The structure and function of blood cells are summarised in Table 5.

Table 5 A summary of the structure and function of different blood cells

Cell type	Structure	Function
Red blood cells	Small (diameter 7–8 µm) cells lacking a nucleus and organelles, and with a biconcave disc shape; packed with haemoglobin.	Adapted to carry oxygen: lack of a nucleus provides more space for haemoglobin; biconcave disc shape increases surface area for gas exchange. Small size facilitates movement through capillaries.
Polymorphs	Cells (diameter 10–12 µm) with a multilobed nucleus and granular cytoplasm. The most common white blood cell (70%).	These short-lived, but numerous cells can squeeze between endothelial cells of capillaries at sites of infection where they engulf bacteria and other foreign bodies by phagocytosis.
Monocytes	Large cells (diameter 14–17 µm) with a kidney-shaped nucleus. The least common white blood cell (5%).	These reside in blood for only a few days before moving out into body tissues, where they develop into macrophages, which are long-lived phagocytic cells that engulf bacteria and foreign material.
Lymphocytes	Cells (diameter 7–8 µm) with a huge nucleus and little cytoplasm. Present in relatively large numbers (25% of white blood cells).	There are two types of lymphocyte: ■ B lymphocytes are involved in antibody production — antibody-mediated immunity ■ T lymphocytes destroy infected cells and foreign tissue — cell-mediated immunity
Platelets	Essentially cell fragments too small to be readily visible using a light microscope.	Have an important role in initiating blood clotting and in plugging breaks in blood vessels.

Plasma

Plasma consists of 90% water and 10% of a variety of substances in solution and suspension: plasma proteins including prothrombin and fibrinogen (involved in clotting), albumin and enzymes, hormones, glucose, amino acids, fats and fatty acids, urea (excretory product), vitamins and various ions (e.g. Ca^{2+} as a clotting factor).

Blood clotting

The clotting of blood seals cuts and wounds, preventing the entry of pathogens and stopping blood loss. When blood vessels are injured, a chain of reactions is initiated (involving clotting factors, such as calcium ions, vitamin K, and factors VIII and Xa). In the final stages of the process, an inactive plasma protein, **prothrombin**, is converted into active **thrombin**. Thrombin is an enzyme that converts soluble **fibrinogen** into insoluble **fibrin**. The fibrin forms a mesh that traps red blood cells and forms a **clot**.

Knowledge check 32

State where the red blood cells 'pick up' oxygen and where they are likely to release it again.

Knowledge check 33

How is a red blood cell adapted to carry haemoglobin and facilitate its functioning in oxygen transport?

Knowledge check 34

Which white blood cells are involved in phagocytosis? How do they differ in activity?

Knowledge check 35

In what form is carbohydrate transported in plants and animals?

A summary of the sequence of events in clotting is given in Figure 27.

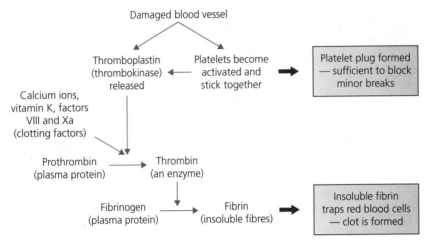

Figure 27 Blood clotting

Knowledge check 36

Name two enzymes involved in clotting and describe their action.

Exam tip

You should be aware that if any of the required factors is missing then clotting is inhibited. For example, removal of calcium ions is used to prevent blood donated by another individual from clotting; if factor VIII is missing, as in the hereditary disorder haemophilia, then clotting time increases.

Haemoglobin and the carriage of oxygen

Haemoglobin is a conjugated protein found in large quantities in red blood cells. Each molecule consists of four polypeptides: two α-chains and two β-chains. Each polypeptide has a **haem** group attached, which contains **iron** (Fe^{2+}). An oxygen molecule can associate with each haem to form oxyhaemoglobin:

$$Hb \quad + \quad 4O_2 \quad \rightleftharpoons \quad HbO_8$$
$$\text{haemoglobin} \quad + \quad \text{oxygen} \quad \rightleftharpoons \quad \text{oxyhaemoglobin}$$

The equation shows that one molecule of oxyhaemoglobin can carry up to four molecules of oxygen. It also shows that the reaction is reversible: if oxygen is plentiful, oxyhaemoglobin is produced; if oxygen levels are low, oxyhaemoglobin releases its oxygen (**dissociates**).

Blood contains a vast number of haemoglobin molecules — each mm^3 of blood contains about 5 million red blood cells and each red blood cell contains about 280 million molecules of haemoglobin. The amount of oxygen carried by all this haemoglobin is measured as the degree to which the blood is **saturated** — 50% saturated means that, on average, each haemoglobin molecule is carrying two oxygen molecules. The amount of oxygen carried by the blood depends on the amount of oxygen available in its surroundings. This is measured as its **partial pressure** (abbreviated pO_2), which is the proportion of the total air pressure that is contributed by the oxygen in the mixture. It is measured in kilopascals (kPa). Figure 28 shows the **oxygen dissociation curve** of human haemoglobin. This indicates how the percentage saturation of blood with oxygen changes with the partial pressure of oxygen.

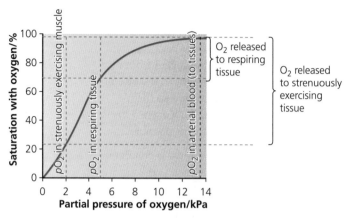

Figure 28 The oxygen dissociation curve of human haemoglobin

While the data for the graph are obtained experimentally, it provides evidence for what is happening in the body. The partial pressure of oxygen in the alveoli is relatively high (13.7 kPa). Therefore, since blood entering the pulmonary circulation is deoxygenated, the haemoglobin **loads** with oxygen to become 98% saturated. In respiring tissues, oxygen is being used up and so the pO_2 is low (5 kPa). At this low pO_2, blood can only be 70% saturated and, since blood arriving at the tissues is highly saturated, haemoglobin **unloads** oxygen to the tissues. During strenuous exercise the pO_2 in muscles is reduced to 2 kPa, at which level blood can only be 24% saturated, and so even more oxygen is unloaded. (Use the graph in Figure 28 to confirm these numbers.)

A significant feature of the graph in Figure 28 is that it is S-shaped (**sigmoid**). This is because of the way the four haem-containing polypeptides interact. When the first oxygen molecule combines with the first haem group, the shape of the haemoglobin molecule becomes distorted. This makes it easier for the other three oxygen molecules to bind with the other haem groups.

The amount of oxygen carried by haemoglobin depends not only on the partial pressure of oxygen but also on the **partial pressure of carbon dioxide (pCO_2)**. The effect of carbon dioxide on the oxygen dissociation curve of human haemoglobin is shown in Figure 29. This shows that at higher carbon dioxide partial pressures, the oxygen dissociation curve moves to the right. This is known as the **Bohr effect**.

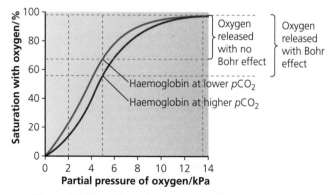

Figure 29 The effect of carbon dioxide on the oxygen dissociation curve of human haemoglobin

The Bohr effect increases the efficiency of haemoglobin for oxygen transport. The partial pressure of carbon dioxide is high in tissues that are actively respiring (e.g. muscle). The higher the level of carbon dioxide, the lower the affinity of haemoglobin for oxygen, and so more oxygen is released to the tissues. This is advantageous because tissues with a high respiration rate require increased amounts of oxygen. In the alveoli the partial pressure of carbon dioxide is low (since it is breathed out), and so the affinity of haemoglobin for oxygen is increased and more oxygen can be loaded. This is advantageous because it allows the blood to become as fully saturated as possible.

Higher temperature and *lower pH* also have a Bohr effect: exercising muscle generates more heat and, as the temperature in the muscle increases, more oxygen is released; as pH is reduced (due to release of CO_2) more oxygen is unloaded.

Figure 30 shows dissociation curves for fetal and adult haemoglobin. Notice that fetal haemoglobin has a higher affinity for oxygen than adult haemoglobin.

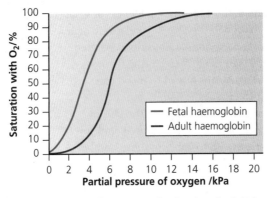

Figure 30 Oxygen dissociation curves for fetal and adult haemoglobin

Myoglobin is found in 'red' muscle. It consists of one polypeptide with a single haem group, and does not, therefore, have a sigmoid dissociation curve. Myoglobin has a very high affinity for oxygen (Figure 31).

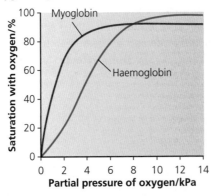

Figure 31 The oxygen dissociation curve of myoglobin compared with that of haemoglobin

Myoglobin only releases oxygen when the partial pressure of oxygen becomes very low. It acts as an **oxygen store** within the muscle. If, as a result of strenuous exercise, the pO_2 becomes very low then myoglobin gives up its oxygen. This enables aerobic respiration to continue for longer and delays the onset of anaerobic respiration.

Knowledge check 38

Explain why fetal haemoglobin must have a higher affinity for oxygen than adult haemoglobin.

Oxyhaemoglobin dissociation curves for different species show considerable differences. For example, the oxygen dissociation curve for llama haemoglobin shows it to have a particularly high affinity for oxygen (Figure 32).

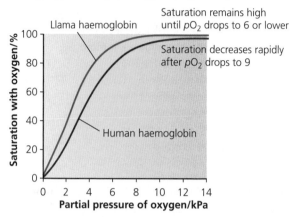

Figure 32 The oxygen dissociation curve of llama haemoglobin compared with that of human haemoglobin

The llama, a mammal of the camel family, is adapted to living at altitudes of about 5000 metres in the Andes mountains of South America. At high altitude, atmospheric pressure is low. The atmospheric pressure at sea level is 101.3 kPa so, with 21% oxygen present, the pO_2 in the air is about 21 kPa, though less, at 14 kPa, in the alveoli. At an altitude of 5500 metres the atmospheric pressure is halved. This means that pO_2 is reduced to 10.5 kPa and the alveolar pO_2 is only 7 kPa. Llama haemoglobin has a *high oxygen affinity* so that the blood can become fully saturated with oxygen at lower pO_2.

Human haemoglobin is not fully saturated at high altitude and an unacclimatised person would begin to show signs of lack of oxygen — breathlessness, nausea and fatigue (mountain sickness). However, a person who moves to high altitude gradually will, after 2–3 days, begin to become **acclimatised**. One of the most obvious changes is an increased production of red blood cells. Thus, there is a greater capacity for carrying oxygen to compensate for the lower levels of oxygen available. This response has been used in training by endurance athletes to increase their potential for oxygen delivery to the muscles when they compete subsequently at low altitude.

Cardiovascular disease

Atherosclerosis

Atherosclerosis (also known as 'hardening of the arteries') is a disease in which an artery wall thickens so that its lumen is narrowed. The process involves the following sequence of events:

1 The endothelium lining the artery gets damaged. This damage can result from **high blood pressure**, which puts an extra strain on the layer of cells, or it might result from toxins from tobacco smoke in the bloodstream.

2 Once the endothelium is breached, macrophages (derived from monocytes) leave the blood vessel and move into the artery wall. These cells accumulate chemicals from the blood, particularly **cholesterol**. A deposit builds up, which is called an **atheroma**.

3 Over time, calcium salts and fibrous tissue build up around the cholesterol, resulting in a hard swelling called a **plaque** (Figure 33). The build-up of fibrous tissue means that the artery wall loses some of its elasticity — the artery is said to 'harden'.

4 Plaques cause a narrowing of the artery lumen so that it is more difficult for blood to flow through it. Blood pressure increases. Raised blood pressure makes it more likely that further plaques will form.

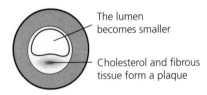

The lumen becomes smaller

Cholesterol and fibrous tissue form a plaque

Figure 33 A plaque in an artery wall

Risk factors include smoking, inactivity, obesity, stress, excessive salt intake, high blood cholesterol, excessive alcohol consumption and diabetes. Risk factors that cannot be controlled include age and **genetic predisposition**.

Coronary thrombosis

Blood platelets tend to collect at the damaged surface of an artery and release factors that trigger blood clotting. A clot that forms inside a damaged blood vessel is called a **thrombus** and the condition is called **thrombosis**. If this happens in the **coronary arteries** it is called a **coronary thrombosis**. The heart muscle supplied by these arteries does not receive any oxygen. If the affected muscle cells are starved of oxygen for long they will be damaged permanently. This results in a heart attack or **myocardial infarction**. If a small branch of an artery is blocked, only a small amount of muscle dies, causing a small heart attack; if a large artery is blocked, the whole heart may stop beating — a **cardiac arrest** (Figure 34).

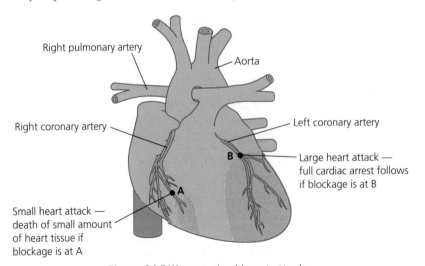

Right pulmonary artery

Aorta

Right coronary artery

Left coronary artery

B ● Large heart attack — full cardiac arrest follows if blockage is at B

● A

Small heart attack — death of small amount of heart tissue if blockage is at A

Figure 34 Different-sized heart attacks

Genetic predisposition Possessing genes that increase an individual's susceptibility to a certain disease.

Knowledge check 39

It may soon be possible to have your DNA analysed to see if you have genes that increase the risk of cardiovascular disease. Suggest a benefit and a potential problem of this.

Myocardial Occurring in heart muscle.

Infarction The death of tissue resulting from oxygen deprivation.

Aneurysm

Atherosclerosis can lead to weakening of a section of the artery wall which, due to the pressure of the blood, bulges outward, forming a balloon-like sac in the artery called an **aneurysm** (Figure 35). If the weakening is severe enough, the aneurysm may burst, causing massive bleeding. Blood loss from a large aneurysm in the aorta is nearly always fatal; blood loss from even a small aneurysm in the brain can put pressure on the brain and cause a stroke.

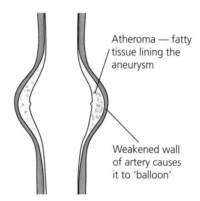

Atheroma — fatty tissue lining the aneurysm

Weakened wall of artery causes it to 'balloon'

Figure 35 An aneurysm in an artery

Diagnosis of cardiovascular disease

In **angiography**, a special dye is released into an artery, making the blood vessels visible when an X-ray is taken. The resultant angiogram shows the blood flow through the blood vessels and allows various blood flow abnormalities to be detected:

- aneurysm, where a section of a blood vessel wall bulges due to a weakness in the wall
- atherosclerosis, where the flow of blood in an artery is restricted or blocked; in coronary angiography, the flow of blood to the heart muscle is assessed to determine which of the coronary arteries are blocked and where

Irregularities in the heartbeat can be detected in an electrocardiogram (ECG). In a normal ECG the wave of excitation, as shown in Figure 21 on p. 29, is repeated at regular intervals from 60 to 100 per minute. An abnormal heartbeat, called an arrhythmia, can involve a change in rhythm, producing an uneven heartbeat, or a change in rate, causing a very slow or a very fast heartbeat.

Diagnosis The identification of a medical condition by examination of the symptoms.

Practical work

Dissection of the mammalian heart:
- identify heart chambers, atrioventricular valves, semilunar valves, chordae tendinae, papillary muscles, interventricular septum, major blood vessels (vena cavae, pulmonary arteries and aorta)

Examine prepared slides and/or photographs of blood vessels (in section):
- distinguish between arteries, veins and capillaries

Examine stained blood films using the light microscope and/or photographs:
- identify red blood cells, polymorphs, monocytes, lymphocytes and platelets

Summary

- Mammals have a double circulatory system: a pulmonary circulation and a systemic circulation.
- There are three stages to the cardiac cycle:
  - diastole, when all heart muscle is relaxed, with blood returning to the atria and flowing into the ventricles
  - atrial systole, when the atrial muscle contracts to 'top-up' the ventricles
  - ventricular systole, when ventricular muscle contracts to force blood into the major arteries
- The heart is myogenic. Excitation originates at the SAN (the pacemaker) and spreads to the AVN before passing down the bundle of His and up the walls of the ventricles via Purkinje fibres.
- The atrioventricular valves are forced open when atrial pressure exceeds that in the ventricles and forced closed when pressure difference is reversed. The semilunar valves are forced closed when arterial pressure exceeds that in the ventricles and forced open when ventricular pressure exceeds that in the major arteries.
- Arteries are adapted to carry blood away from the heart, under high pressure, to the organs. Veins are adapted to return blood, under low pressure, to the heart.
- Exchange takes place in the capillaries, with tissue fluid carrying substances (e.g. oxygen and glucose) to the cells.
- Blood consists of plasma and cells:
  - red blood cells to carry haemoglobin
  - lymphocytes to form B and T cells in the immune response to antigens
  - polymorphs and macrophages (formed from monocytes) to carry out phagocytosis
- Haemoglobin is adapted to 'pick up' oxygen at high pO_2 (in lungs) and release it at low pO_2 (in respiring tissue). Further dissociation of oxygen takes place if the pCO_2 is increased as a result of increased respiration.
- Blood clotting seals damage to vessels. It involves platelets and the plasma proteins, prothrombin and fibrinogen.
- Atherosclerosis (thickening of arterial walls) and coronary thrombosis (clotting within coronary arteries) are examples of cardiovascular disease.
- Angiograms allow the diagnosis of atherosclerosis, thrombosis and aneurysm.

■ Adaptation of organisms

Organisms live in a **habitat**. They are part of an **ecosystem** within which they interact with both their **biotic** environment (the other living organisms) and the **abiotic** environment (physical and chemical factors). Members of a single species form a **population** and, with other populations, make up a **community**.

Adaptations

Each organism has an **ecological niche** that describes its role within the ecosystem:

- what it feeds on or other nutrient needs
- what feeds on it
- competition with other organisms
- its temperature, water and other requirements

Students often have difficulty describing the ecological niche of an organism. What is required is a full description of the environmental requirements for it to survive and reproduce. For example, a fox is a carnivore that hunts mostly at night. However, it may also scavenge; it may compete with other predators (e.g. stoats); it is not predated, though human activities are a major threat (e.g. road deaths); its habitat includes woodland edges where it may dig an 'earth'.

If it is to survive and reproduce, an organism must have adaptations for each of the above points. Adaptations may be behavioural, physiological or morphological. A **behavioural adaptation** is an aspect of the behaviour of an organism that helps it to survive and reproduce — for example, in attracting a mate. A **physiological** or **biochemical adaptation** is one in which there is appropriate functioning of the organism or its cellular processes — for example, the ability to respire anaerobically. A **morphological** or **anatomical adaptation** refers to any structure that enhances the survival of an organism — for example, the spines on a cactus that prevent grazing.

Adaptations of hydrophytes — plants specifically adapted for living in water — are shown in Figure 8 on p. 14.

Adaptations of xerophytes — plants adapted to living in dry habitats — are summarised in Table 3, and illustrated in Figure 17 on p. 25. This can be further expanded upon:

- Some xerophytes close their stomata when little water is available and some only open their stomata at night, when transpirational loss is reduced. These are behavioural adaptations.
- Some xerophytes possess cells that store water when it is readily available — the plants have succulent leaves and/or stems — for use in times of shortage. This is a physiological adaptation.
- Xerophytes have many morphological or anatomical adaptations such as sunken stomata that trap a layer of moist air next to the stomata, reducing the water potential gradient for water vapour to diffuse into the atmosphere.

Distribution of organisms

The distribution of organisms within their ecosystem is influenced by biotic factors and abiotic (climatic and edaphic) factors.

Climatic factors include:

- **Temperature range** The Sun is the main source of heat for ecosystems. The temperature range within which life exists is relatively small. At low temperatures ice crystals may form within cells, causing physical disruption; at high temperatures enzymes are denatured. Fluctuation in environmental temperature is more extreme in terrestrial habitats than aquatic ones because the high heat capacity of water effectively buffers the temperature changes in aquatic habitats.
- **Availability of water** The availability of water is a main factor in determining the distribution of terrestrial organisms. Many have adaptations to conserve water (e.g. the waxy cuticle of flowering plants and insects); some have less effective

waterproofing and so are confined to moist or humid localities (e.g. mosses and woodlice). Even in aquatic habitats there may be problems due to the osmotic movement of water. In marine ecosystems, fish (e.g. mackerel) have adaptations to conserve water since there is a tendency for water to be withdrawn osmotically. Some fish (e.g. perch) live exclusively in freshwater and have adaptations to reduce osmotic gain. A few fish (e.g. salmon) are capable of tolerating both extremes during their life cycles.

- **Light intensity** As the ultimate source of energy for ecosystems, light is a fundamental necessity. All plants require light for photosynthesis and most grow better the more light they receive. Some plants, however, are tolerant of low-light conditions, while others (e.g. bluebells) are adapted to grow on the woodland floor before the leaf canopy develops above. In aquatic ecosystems light is absorbed by the water molecules, so aquatic plants and algae can photosynthesise effectively only if they are near the surface. There is even less light in aquatic systems containing suspended particulates such as organic matter.

- **Light quality** Water not only absorbs light, it also influences which wavelengths can penetrate. Blue light penetrates water to a greater depth because red light is absorbed. Some marine algae — the red seaweeds — possess additional red pigments specifically to absorb at the blue end of the spectrum, and so red algae are adapted to live at greater depths than most other algae.

- **Day length** The longer the day length the more time a plant has for photosynthesis. The greater growth of plants in summer is more to do with the increased day length than it is to do with a rise in temperature.

Edaphic (soil) factors include:

- **pH values** The pH of a soil influences the availability of certain ions. Plants such as heathers grow best in acid soils, while spring gentian and cowslip prefer alkaline soils. Species that are tolerant to extremes of pH can become dominant in particular areas because competing species cannot survive in these extreme conditions. The dominance of heathers on upland moors is, in part, due to their ability to withstand very low soil pH. However, the optimum pH for the growth of most plants is close to neutral.

- **Availability of nutrients** There is a wide variety of ions required by plants. Some of these are needed in relatively large amounts and are called macronutrients: nitrate for amino acids synthesis, phosphate for nucleotide synthesis, calcium for the middle lamella, sulfate for the synthesis of some amino acids and iron for the production of chlorophyll. Some ions are required in minute amounts and are called micronutrients. Different species make different demands on the ions in the soil and therefore plant distribution depends to some extent on the nutrient balance of a particular soil.

- **Water content** The water content of soils varies markedly. It depends on the soil type — clay soils tend to hold a lot of water, sandy soils are freely draining and hold little. A waterlogged soil creates anaerobic conditions. Plants able to tolerate these conditions include the rushes (*Juncus* spp.) and sedges (*Carex* spp.). They have air spaces within their root tissues that allow some diffusion of oxygen from the aerial parts to help supply the roots.

Exam tip

Environmental conditions may vary gradually over a relatively short distance, resulting in **zonation**. This can be seen on a rocky shore where some species of seaweed, resistant to exposure, are found on the upper shore, while other, less tolerant species are confined to the lower shore where they are more often submerged in water.

Knowledge check 41

What are edaphic factors?

■ **Aeration of soils** The space between soil particles is filled with air, from which the roots obtain their respiratory oxygen by diffusion. Soil air is also necessary to the aerobic microorganisms in the soil that decompose the humus.

Biotic factors include:

■ **Competitors** Organisms compete with one another when they share a common resource (e.g. food, water, light or ions), and that resource is in limited supply. They compete not only with members of other species — **interspecific competition** — but also with members of their own species — **intraspecific competition**. Where two species occupy the same ecological niche, the interspecific competition leads to the local extinction of one or the other — the **competitive exclusion principle**.

■ **Predators** The distribution of a predator is reliant on the presence of its prey species. The population numbers of both prey and predator are interdependent — when prey numbers are low, predator numbers decline, and when predator numbers become high (due to abundant prey), then prey numbers drop.

■ **Accumulation of waste** The growth of microorganisms is frequently self-limiting because the accumulation of waste products can be toxic — for example, in anaerobic conditions yeast populations produce ethanol.

Practical work

Describe and carry out qualitative and quantitative techniques used to investigate the distribution and relative abundance of plants and animals in a habitat:
■ Sampling procedures to include:
  – random sampling
  – line transect
  – belt transect
■ Sampling devices to include quadrats, pin frames, pitfall traps, sweep nets and pooters.
■ Estimation of species abundance, density, frequency and percentage cover.
■ Appreciation and, where possible, measurement of the biotic and abiotic factors that may be influencing the distribution of organisms.

Summary

■ Each species of organism has an ecological niche — a set of environmental conditions in which it can survive and reproduce.
■ Adaptations within an environment may be behavioural, physiological or morphological.
■ Hydrophytes are flowering plants adapted for living in water, while xerophytes are plants adapted for living in dry habitats.

■ Environmental factors affect the distribution of organisms and include abiotic — climatic and edaphic (soil) factors — and biotic factors.
■ Interspecific competition is an important biotic factor. The competitive exclusion principle is that no two species can occupy the same ecological niche.

■ Diversity of life

Classification

There is such a huge diversity of life on Earth that it makes sense to try to provide some degree of order by categorising the organisms — putting them into groups according to their similarities and differences. This grouping of organisms is known as **classification** and the study of biological classification is called **taxonomy**. Biological classification attempts to classify living organisms according to how closely related they are. It makes use of information from all areas of biology — for example, cell structure, immunology, physiology, anatomy, behaviour, life cycles, ecology — and also from biochemistry. The basic unit of biological classification is the **species**.

The concept of the species

A species is defined as 'a group of organisms that is capable of interbreeding to produce viable and fertile offspring'.

This definition attempts to take into account some of the difficulties in describing 'what a species is':

- Robins in Ireland do not interbreed with robins in France, because they are geographically isolated, but they are *capable* of doing so — they are members of the same species.
- A horse and a donkey are capable of interbreeding, but their progeny, the mule, is *sterile* — they are different species.

Members of the same species have numerous features in common while still exhibiting some degree of variation. They have many genes in common, but variation is generated by the different alleles.

Classifying species

Species are classified into groups or categories of increasing size, i.e. **species**, **genus**, **family**, **order**, **class**, **phylum** and **kingdom**. Each of these groups is called a **taxon** (plural **taxa**) — Table 6.

Table 6 The hierarchy of taxa

Taxon	Description
Genus (plural genera)	Group of related species
Family	Group of related genera
Order	Group of related families
Class	Group of related orders
Phylum (plural phyla)	Group of related classes
Kingdom	Group of related phyla

The Latin names of all taxa except those of species take initial upper case letters, while anglicised versions do not — for example, the kingdom Animalia has members that are called animals.

> **Knowledge check 42**
>
> A biologist wished to determine whether the members of two populations belonged to the same species. How could this be achieved?

> **Exam tip**
>
> Be prepared to place taxa in the correct order. The descending order of **k**ingdom, **p**hylum, **c**lass, **o**rder, family, **g**enus and **s**pecies can be remembered as 'keep pond clean or frogs get sick' (kpcofgs).

Naming species

Species are named using their genus and species. This is called the binomial system. By international convention, genus and species names are written in italics (underlined when handwritten); the genus name has an upper case initial letter and the species has a lower case initial letter — for example, the European robin is named *Erithacus rubecula*. When a binomial has been used once in a passage it may be shortened subsequently (e.g. *E. rubecula*), providing that this does not cause confusion. If the species is not certain but the genus is known, it is possible to refer to an organism by the name of the genus followed by 'sp.' (plural 'spp.') — for example, *Lumbricus* sp., a species of earthworm.

The five kingdoms

Until quite recently, biologists classified all living organisms into five kingdoms (Figure 36). These five kingdoms are:

- **Prokaryotae**
- **Protoctista**
- **Fungi**
- **Plantae**
- **Animalia**

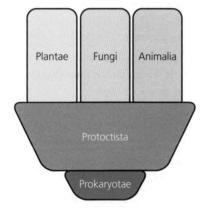

Figure 36 The five-kingdom classification

Members of each kingdom have features in common, though some of these are also possessed by members of other kingdoms. For example, members of all the kingdoms except the Prokaryotae have eukaryotic cells. So, while having eukaryotic cells is a feature of the kingdom Plantae, it is *not* a distinguishing feature; members of the kingdom Plantae are distinguished by the possession of cellulose cell walls. Table 7 summarises the features of each of the five kingdoms, with distinguishing features shown in bold.

Exam tip

The name given to each species is *both* names in the binomial. The second name should *not* be called the species name. Rather, it is called the specific name, unique to a species within the genus.

Knowledge check 43

Why is it important for scientists to use Latin names, rather than common names, for organisms mentioned in research papers?

Table 7 The features of members of each of the five kingdoms

Kingdom	Features (distinguishing features in bold)
Prokaryotae (prokaryotes) Examples include bacteria, blue-green algae (cyanobacteria) and hot-spring bacteria (belonging to the Archaea domain)	**Cells are microscopic and prokaryotic: they lack a nucleus and membrane-bound organelles; the DNA is circular** (and not histone-bound in bacteria); **ribosomes are smaller than those of eukaryotes**; cell walls are made of peptidoglycan in bacteria (but not in archaebacteria). **Since nuclei are lacking, cell division occurs by simple binary fission.** Different methods of nutrition are exhibited by the prokaryotes.
Protoctista (protoctistans) Examples include *Pleurococcus* (a green alga), which is autotrophic, and *Paramecium* (a ciliated protozoan), which is heterotrophic	This is a diverse group and membership is often by exclusion from all other groups. Some are unicellular, some are filamentous while others are multicellular, though show limited differentiation; all protoctistans have eukaryotic cells. Some possess cell walls (cellulose or non-cellulose) and chlorophyll (enabling photosynthesis), while others have no cell walls and are motile; they can be autotrophic or heterotrophic.
Fungi (fungi) Examples include yeast, moulds (e.g. the bread mould, *Rhizopus*), mushrooms and toadstools	Fungi are most often multicellular, though a few are unicellular (e.g. yeast). Cells are eukaryotic (most often organised into filaments, or hyphae, and frequently multinucleate and not divided into separate cells); and **cells have a cell wall, often made of chitin**. **Fungi have a lysotrophic method of nutrition: they secrete enzymes to digest organic materials (usually dead) outside their cells (extracellular digestion) and absorb the products of digestion**; they are important decomposers, breaking down and recycling organic matter. They store carbohydrates as glycogen, and lipids as oils.
Plantae (plants) Examples include mosses, ferns, conifers and flowering plants	Multicellular; cells are eukaryotic; **cells possess a cellulose cell wall**. **All plants are photosynthetic, containing chlorophyll in chloroplasts, and are autotrophic.** **They store carbohydrates as starch**; and lipids as oils.
Animalia (animals) Examples include flatworms, segmented worms, arthropods (e.g. insects) and chordates (fish, amphibians, reptiles, birds and mammals)	Multicellular; cells are eukaryotic; **cells lack a cell wall**. **All animals are heterotrophic; most ingest food into a digestive system**. They store carbohydrates as glycogen; and **usually store lipids as fats**. **Most are capable of locomotion.**

Autotrophic Producing its own food (carbohydrates, fats and proteins) from simple inorganic substances, generally using light energy in photosynthesis.

Heterotrophic Obtaining food by digestion of complex organic compounds (polysaccharides, fats and proteins); lysotrophic is one type of heterotrophic nutrition.

Knowledge check 44

Describe the differences between plants and fungi.

Knowledge check 45

Name one characteristic in each of the following kingdoms:
a fungi
b animals

Knowledge check 46

Apart from being eukaryotic, identify two features that fungi and animals have in common.

The three domains

Biological classification schemes are devised by taxonomists, based on the best available evidence at the time. New information, in 1990, giving greater weight to findings from molecular biology, has led to the proposal of a completely new classification. As a result of this information, three forms of life, called **domains** (superkingdoms), are recognised. The prokaryotes are now classified as two domains, the **Archaea** and the **Bacteria**. All other organisms (i.e. the eukaryotes) are classified within the domain **Eukarya** (Figure 37).

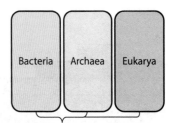

Figure 37 The three-domain classification

The basis for the three-domain system is, in the main, due to unique differences in the RNA molecules in ribosomes. Nevertheless, there are other distinguishing features:

- Bacteria and Archaea lack membrane-bound organelles and have a circular DNA molecule.
- Archaea and Eukarya lack a peptidoglycan cell wall (though many Eukarya have no cell wall) and have histones (proteins) bound to the DNA.
- Archaea membranes contain phospholipids that differ from those of Bacteria and Eukarya.

A phylogenetic classification

The most natural system of classification is one that reflects the ancestral or evolutionary relationships between groups. This system of classification is called **phylogenetic**. It puts the more closely related organisms together into the smaller groups, i.e. genus, then family, etc. Thus, the genus *Panthera* includes the closely related lion (*P. leo*) and tiger (*P. tigris*) but excludes the cheetah (*Acinonyx jubatus*) since it has unique features, although it possesses sufficient similarities to be included within the same family, Felidae.

Usually, the more features in common two species have, the more recently they have evolved from a common ancestor.

The sequencing of the amino acids in proteins and the nucleotides in DNA and RNA molecules allows organisms to be compared at the most basic level — the gene. Closely related organisms possess a high degree of agreement in the molecular structure of their DNA, RNA and protein; molecules of organisms related distantly usually show dissimilarity.

Analysis of amino acid differences in proteins (e.g. cytochrome c)

Protein analysis is used to compare the amino acid sequences of the same protein in different organisms. The more differences, the less closely related the species. Proteins that have been analysed include haemoglobin and cytochrome c, a protein of about 100 amino acids involved in aerobic respiration. Figure 38 shows the number of amino acid differences in the cytochrome c of six organisms and the phylogeny based on the cytochrome c data. While a classification cannot reliably be based on the analysis of a single protein, the results conform well to phylogenies based on other features.

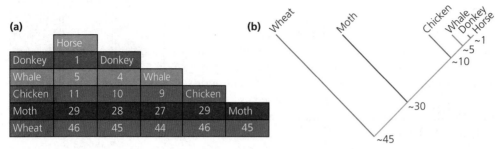

Figure 38 Cytochrome c analysis showing (a) the number of amino acid differences among six organisms, and (b) the resultant phylogeny

Analysing protein differences using immunological techniques

The protein albumin (consisting of 584 amino acids) is found in the blood plasma of many animals, although different species have differences in the amino acid sequence. Without having to obtain the sequences, the differences can be measured using an immunological technique. This was used to measure how similar human albumin is to albumin from the great apes and from an Old World monkey, the baboon (Figure 39). Human albumin was injected into rabbits to produce anti-HA (human albumin) antibodies; when anti-HA antibodies were mixed with human albumin there was 100% precipitation; when anti-HA antibodies were mixed with albumin from other primates there was a lesser degree of precipitation according to the number of amino acid differences from human albumin. The greater the degree of precipitation, the fewer amino acid differences in albumin and the closer the evolutionary relationship of that primate to humans.

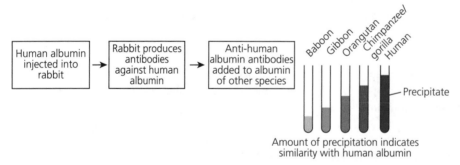

Figure 39 Immunological comparisons of human albumin with that of the great apes

Using this immunological technique, a phylogenetic tree was developed (Figure 40).

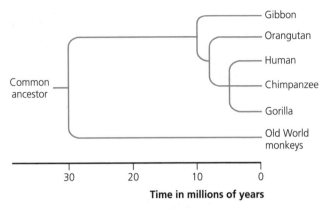

Figure 40 Phylogenetic tree for humans and great apes based on immunological data

Analysing differences in DNA sequences using DNA hybridisation

Comparing DNA sequences is the most accurate way of demonstrating relatedness. The greater the similarity between the sequences, the more closely related the two species are. The technique of DNA hybridisation involves extracting samples of DNA from two species, separating the DNA into single strands, mixing these strands and allowing them to re-bind (hybridise). Hybridised strands with a high degree of similarity will re-bind more firmly (Figure 41).

Knowledge check 47

DNA from two species was hybridised. The two strands separated at very low temperatures. Explain what this tells us about the two species.

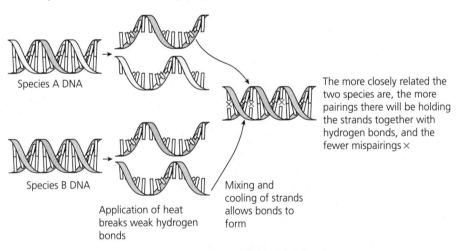

Figure 41 Technique of DNA hybridisation

Samples of DNA from closely related species show a high degree of hybridisation; it is lower for more distantly related species. Using DNA hybridisation techniques, the phylogenetic tree of humans and the great apes appears as in Figure 42.

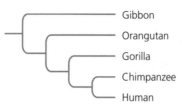

Figure 42 The phylogenetic tree
of humans and the great apes,
based on DNA hybridisation data

This is different from, and arguably more accurate than, the tree shown in Figure 40.

Analysing differences in genes using DNA sequencing

DNA sequencing determines the exact order of nucleotides (bases) in a DNA molecule. The process is used to directly compare the base sequences of a gene found in different species. For example, it is possible to compare the nucleotide sequences of the human cytochrome c oxidase gene with that of the ape species, in order to determine the degree of relatedness between the species.

Skills development

Construction of a table to display results

The most convenient way of recording results is in the form of a table. Careful consideration must be given to how the data are organised. The arrangement should make it easier to identify any trends and draw conclusions (Table 8).

Table 8 Important steps in the construction of a table

Tabular feature	Description of process
The table should have a caption (also called a legend or a title).	This must include the independent variable (e.g. the effect of X on...), the dependent variable (the process being affected) and the biological material being investigated.
There should be a suitable number of columns, each with an appropriate heading.	The first column is for the independent variable (i.e. the variable that you changed in the experiment). Subsequent column(s) is/are for the dependent variable (i.e. what you measured or counted).
The column headings should include the units of measurement (as appropriate).	The heading is followed by a solidus (or slash) and then the unit of measurement. SI units should be used.
There should be an appropriate number of rows.	This should include the rows for headings as well as rows to cover the range of the independent variable.
The data should be added in a uniform way.	They should be consistent in the number of significant figures (often three significant figures, e.g. 6.34). Very large or very small numbers are more easily expressed in standard form (e.g. 3 750 000 as 3.75×10^6 or 0.000271 as 2.71×10^{-4}).

Table 9 shows a table prepared to record the results from an investigation in which the effects of two types of amylase are studied over a range of temperatures.

> **Exam tip**
>
> You are not expected to know the techniques involved in protein and DNA analysis. However, you are expected to understand the principles and to be able to apply this understanding in answering examination questions.

→

Table 9 The effect of temperature on the digestion of starch by two amylase enzymes

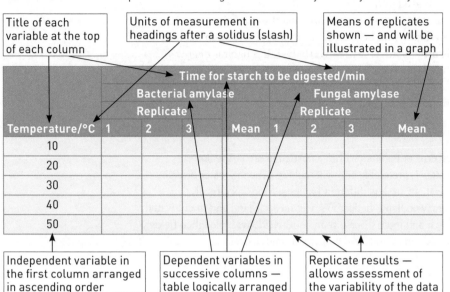

Title of each variable at the top of each column

Units of measurement in headings after a solidus (slash)

Means of replicates shown — and will be illustrated in a graph

Independent variable in the first column arranged in ascending order

Dependent variables in successive columns — table logically arranged

Replicate results — allows assessment of the variability of the data

Temperature/°C	Time for starch to be digested/min							
	Bacterial amylase				Fungal amylase			
	Replicate				Replicate			
	1	2	3	Mean	1	2	3	Mean
10								
20								
30								
40								
50								

Biodiversity

Biodiversity is a measurement of the variety of living organisms within a particular area, and has three components:

- species diversity
- ecosystem (or habitat) diversity
- genetic diversity

Species diversity

Species richness is the number of different species present in an area. This measure has limited use, because a habitat may be species-rich but not show much diversity — only one or two species may be especially abundant, with the majority being sparsely present.

Simpson's index (D) is a more reliable measure of diversity because it takes into account both the number of different species (richness) and the abundance (e.g. number of individuals) of each species present. The formula for calculating D is as follows:

$$D = \frac{\Sigma n_i(n_i - 1)}{N(N - 1)}$$

where Σ = the sum of, n_i = the total number of organisms of each individual species and N = the total number of organisms of all species.

The value of D ranges from 0 to 1. With this index, 0 represents infinite diversity and 1 represents no diversity. That is, the bigger the value of D, the lower the diversity.

To calculate Simpson's diversity index for a particular area, the area must be sampled and the abundance of each species noted. For example, the diversity of the ground flora in woodland might be determined by sampling with random quadrats. Within each quadrat, the abundance each species could be obtained as a number, percentage cover or frequency. However, there must be consistency, i.e. either all by 'number of individuals', or all by 'percentage cover', or all by 'frequency'.

Exam tip

In the exam you could be asked to calculate Simpson's diversity index. The formula for Simpson's diversity index (D) will be provided together with the definitions of n and N. You need to remember that D lies between 0 and 1, with a lower number representing greater biodiversity.

Species biodiversity can be used to indicate the 'biological health' of a particular habitat. However, care should be used in interpreting biodiversity measures. Some habitats (e.g. Arctic tundra) are stressful and so few organisms are adapted for life there, but those that are adapted may well be unique or, indeed, rare. Such habitats are important, even if there is little biodiversity. Nevertheless, if a habitat begins suddenly to lose its animal and plant types, ecologists become worried and search for causes, such as a pollution incident. Alternatively, an increase in the biodiversity of an area may mean that corrective measures have been effective.

Ecosystem diversity

This is the **diversity of ecosystems** or habitats within a particular area. A region possessing a wide variety of habitats is preferable and will include a much greater diversity of species than a region in which there are few different habitats. More specifically, countryside that has ponds, rivers, woodland, hedgerows, wet meadowland and uncultivated grassland will be more species rich and more diverse than countryside with ploughed fields and drained land that is without wet areas and devoid of woods and hedgerows.

Genetic diversity

This is the **genetic variability** of a species. It refers to the variety of alleles possessed by the individuals of a population. Genetic variability allows a population to adapt to changing environments. With more variation, it is more likely that some individuals in the population will possess alleles that are suited for the environment. Lack of genetic diversity is seen as problematic. It indicates that the species may not have sufficient adaptability and may not be able to survive an environmental hazard. The Irish potato blight of 1846, which resulted in the death of a million people and forced another million to emigrate, was the result of planting only two potato varieties, both of which were vulnerable to the potato blight fungus, *Phytophthora infestans*.

Knowledge check 48

Seed banks contain seeds of the ancestors of modern crop plants. Suggest why these seeds are kept.

Summary

- A species (the basic unit of classification) is a group of similar, closely related organisms that are capable of interbreeding with each other to produce viable and fertile offspring.
- Species are classified into groups within a hierarchy. In descending order this hierarchy is: kingdom, phylum, class, order, family, genus, species.
- Species are named according to a binomial system. The genus name is given first, followed by a species name.
- Organisms are grouped into five kingdoms: Prokaryotae (bacteria), Protoctista, Fungi, Plantae, Animalia. Members of the Prokaryotae have prokaryotic cells; members of the other kingdoms have eukaryotic cells. Alternatively, organisms can be classified within three domains: Bacteria, Archaea and Eukarya.

- A phylogenetic classification reflects the ancestral or evolutionary relationship between groups.
- Modern methods of establishing relationships include comparison of amino acid sequences in proteins and of nucleotide sequences in DNA and RNA. Closely related organisms possess a high degree of agreement in the molecular structure of their proteins, DNA and RNA.
- Species diversity can be measured by using the Simpson's diversity index (D). Values for D lie between 0 and 1, with values closer to zero indicating greater diversity.
- The 'richness' of a region (such as Northern Ireland) can be described in terms of the diversity of its ecosystems (or habitats).
- Genetic diversity reflects the genetic variability of a species and is important in allowing for the continued adaptability of a species.

■ Human impact on biodiversity

The impact that humans have on the environment is unlike that of any other species. Areas of particular concern include agriculture and pollution.

The influence of agricultural practices

The human activity that most impacts on the environment is agriculture. While the most basic task of farmers is to provide food for society, they also have control of the countryside environment.

Cultivation methods

High-yield strains (genetic types) of *single-species crops* — **monocultures** — are grown in large areas of land. Since monocultures consistently take the same nutrients from the soil, it soon becomes depleted of particular ions, necessitating the application of fertiliser. Soil is tested for its ion content so that the most suitable fertiliser is prepared. Further, monocultures allow **pests** and diseases to spread rapidly and then build up year after year, so that pest management is required.

The use of monocultures in **intensive farming** has been associated with a huge loss in biodiversity: single-species crops support few other species; fertiliser runoff leads to eutrophication; pesticide use kills beneficial insects. In spite of the difficulties, monocultures are favoured because crop yields can be high, while there is an economic advantage in specialisation (one set of machinery for planting and harvesting a single crop, with reduced labour costs through mechanisation).

> **Exam tip**
> You should be aware that intensive farming practices were introduced in the 1950s (after the Second World War and the rationing of food) to increase agricultural output. From 1940 to 1980, the UK went from 30% to 80% self-sufficient in crops (i.e. a productivity increase of over 160%). However, the huge environmental (and health) costs became unsustainable (some farmland birds declined by over 80%).

Typically, yields in monoculture crops decline, due to a combination of soil-borne plant pathogens and exhaustion of particular soil nutrients. The advantages of **crop rotation**, with *different crops planted in the same field from one year to the next*, then become obvious:

- There is increased soil fertility, because different plant species have different nutrient demands.
- It becomes harder for pests and pathogens to become established, because the same crop species is not grown continuously.

Often, shallow-rooted and deep-rooted plants are used in sequence, for example wheat and then turnips; in one year, legume crops (e.g. clover or peas) may be planted. Also, because of the greater variety of crops grown, crop rotation supports greater species diversity. However, while there are savings in the reduction of fertiliser and pesticide use, there are extra costs (increased labour and different machinery for different crops).

> **Exam tip**
> Even better than testing the soil, is analysing the ion content of the crop plant. For example, in the apple orchards of Co Armagh, leaves are tested to determine their nutrient needs.

Pest In agriculture, any organism that causes a loss in crop yields exceeding 5–10% of total yield.

Intensive farming The growing of high-yield, single-species crops, using fertilisers and pesticides to maximise productivity.

> **Knowledge check 49**
> What are the advantages of crop rotation? What are the disadvantages?

Some farmers favour **polyculture**, for example intercropping, with *two or more crops grown at the same time on the same plot*. Crops may be chosen to mature at different times (e.g. radishes and turnips) or may be planted with a view to their mutual benefit (e.g. a crop that attracts a natural predator of aphids helps a nearby crop that is vulnerable to aphids). Polyculture has a number of advantages:

- Pests and pathogens are unable to spread through the entire crop because different species are grown, and so pest control is easier.
- There is increased soil fertility, because different plant species have different nutrient demands.
- Species diversity is supported.

However, labour costs are greatly increased through planting, harvesting and marketing different crops within the same year.

Hedgerows

Following a previous period of hedgerow removal, farmers are now encouraged to plant new hedges and replant those that have become gappy through neglect (with grant-aid available from the Department of Agriculture and Regional Development, DARD). A mixture of species must be planted and native trees included in the hedgerow. The aim is to include a range of species and so provide a variety of niches for a large diversity of animal species.

Good hedgerow management through trimming also aims to promote biodiversity:

- Maintain a variety of hedge heights and widths to provide the optimum range of habitats (different bird species have different preferences).
- Trim in January/February to avoid destruction of birds' nests (March–August) and allow the berry crop to be used by wintering birds (September–December)
- Trim on a 2- or 3-year rotation rather than annually, to boost the berry crop and insect populations and thereby bird populations.
- Avoid trimming all hedges in the same year (to allow hedge diversity).
- Avoid cutting native hedgerow trees as these support a greater range of insect and bird species.

Uncultivated field margins

Farmers are also encouraged to leave an **uncultivated margin** at field perimeters. The aim is to provide additional habitat and food sources for a range of insects, birds and mammals, so contributing to the recovery of priority species. It should also encourage natural predators, for example some beetles, for pest control; for this reason, such margins are called **predator strips**. There is also support for farmers to plant native trees within the field margins. These strips also extend the wildlife corridors along which many species can move between habitats.

Pest management

Crops that are susceptible to pests and diseases may require the application of pesticides: **herbicides**, **insecticides** or **fungicides**. However, indiscriminate pesticide use has been found to be problematic both in terms of its negative effect on biodiversity and its long-term effectiveness:

Knowledge check 50

Why is the annual cutting of hedgerows detrimental to biodiversity?

Knowledge check 51

Suggest how Simpson's diversity index for a mixed woodland of native trees would compare with that for a conifer plantation.

- Pests have rapid rates of population growth, so **resistant strains** tend to evolve.
- Pesticides may be toxic to species other than the pest species (these are **'broad-spectrum' pesticides**), leading to:
  - **pest resurgence** — greater numbers of pests return because a natural predator is killed
  - **secondary pest outbreak** — a minor pest multiplies rapidly in the absence of its competitor
- Pesticides may **persist** in the environment and adversely affect other ecosystems, for example herbicide spray may drift into adjacent areas and kill plant species important in the food chains of various animals.
- Persistent pesticides may also be **non-biodegradable** (not broken down in animal tissues), so there is a build-up along the food chain, a phenomenon called **bioaccumulation**, reaching toxic levels in predators.

Knowledge check 52

Why might the use of pesticides be ineffective in controlling the numbers of pest species?

Integrated pest management, IPM (Table 10) involves the development of an overall strategy, with a *range of control measures* and the goal of significantly reducing or eliminating the use of pesticides while at the same time managing pest populations at an acceptable level.

Table 10 Features of integrated pest management

IPM measures	Comments
Selecting varieties best suited for local growing conditions	The most productive variety would be avoided if it was susceptible to local pests
Crop rotation and/or intercropping	Harder for pest species to become established
Monitoring pest levels	To determine the need for control measures
Using photodegradable plastic around and between crop plants (e.g. maize)	Prevents the growth of weeds, reducing dependence on herbicides
Using methods to disrupt the breeding of pests	■ Crop rotation ■ Use of mechanical traps ■ Use of sterile males of pest species
Using specific, natural predators or parasites of pests — **biological control**	■ Long-term control without the problems associated with chemical control ■ Emphasis on control, not eradication
Using narrow-spectrum, biodegradable pesticides, which target the pest species; only used as a last resort when the pest may cause economic damage to the crop	■ Greater specificity in killing pest species means that pollinators and natural predators are not affected ■ No bioaccumulation in food chains

Biological control The deliberate introduction of a predator, parasite or pathogen to reduce the pest numbers to a point below that at which it causes economic damage.

Exam tip

Farmers are now more aware of the environmental implications of pesticide application — for example, an oilseed rape grower will only apply a (biodegradable) insecticide when an infestation of aphids has reached economic damage levels, and then only during the hours of darkness when bees (priority species) are inactive.

The pollution of waterways

Lakes and rivers are polluted by the addition of organic matter or inorganic nutrients, because these can adversely affect human health and the survival of other organisms.

Organic pollution

Organic pollutants include sewage, slurry from intensive livestock units, silage effluent and seepage of milk from dairies.

There are bacteria in water that use the organic waste as food. Consequently, if organic waste enters a waterway, there is a population explosion of these decomposing bacteria. The bacteria use up oxygen as they respire, reducing the oxygen level in the water. They place a high oxygen demand upon the water. The quality of a body of water can be measured by its **biological oxygen demand** (**BOD**). The *lower the BOD*, the fewer bacteria are present, indicating that there is *less organic material* in the water; the higher the BOD, the more organic pollution is evident.

Whenever organic waste such as sewage gets into a river, there are changes downstream from the point of discharge. Not only are there changes in the BOD and oxygen levels (Figure 43a), but there is also a succession of different species of aquatic invertebrates (Figure 43b). Specific species are adapted to different levels of oxygen in the river, and thus act as '**indicator species**' of the degree of pollution. The presence of, for example, midge larvae (bloodworms) indicates low levels of oxygen with a high level of pollution; the presence of stonefly and mayfly nymphs indicates high levels of oxygen, so no pollution.

(a)

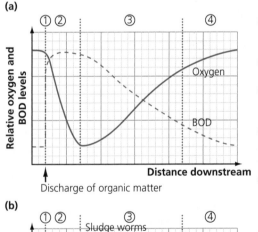

① Initially, the discharge of organic matter provides food for bacteria, which multiply dramatically — there is a high BOD

② The bacteria use up oxygen so that oxygen levels fall

③ As the organic matter is decomposed the oxygen levels start to rise and the BOD falls

④ Eventually, much of the organic matter is used and the oxygen levels return to the normal level

(b)

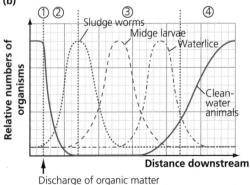

① Clean-water invertebrates, such as stonefly larvae and mayfly nymphs, and fish, such as trout, swim away or are killed since they cannot survive in water with little oxygen

② Only organisms adapted to survive in water with very low levels of oxygen, such as sludge worms (*Tubifex*), can survive, so their numbers increase

③ As oxygen levels gradually rise there is a succession of organisms adapted to lower-than-normal oxygen levels, such as midge larvae (*Chironomus*) and waterlice (*Asellus*)

④ As oxygen levels return to normal, pollution-tolerant species decline due to increased predation and competition from returning pollution-intolerant (clean-water) species

Figure 43 The discharge of organic matter into a river causes changes in (a) oxygen levels and BOD, and (b) the distribution of organisms in the river

DARD sets out management requirements for farm waste (including slurry and silage effluent):

- Farm waste must be stored, to prevent seepage or runoff, until it can be spread onto fields.
- Discharge of farm waste into waterways or in the vicinity of waterways is not permissible.
- Spreading manures onto fields is only permissible at certain times of the year (February–October), and then only when there is no risk of pollution occurring (i.e. not where the land is waterlogged or sloping towards a waterway or when heavy rain is forecast).

Exam tip

Students often confuse organic pollution of waterways with eutrophication. Organic pollution results from the deposition of sewage, slurry or silage effluent into the waterway and causes oxygen depletion as aerobic bacteria digest the organic matter. Eutrophication is nutrient enrichment of waterways and results from the addition of inorganic ions, which may cause algal blooms. There is overlap in what happens subsequently, because dead algae will be digested by aerobic bacteria, causing oxygen depletion.

Eutrophication

Eutrophication is the nutrient enrichment of water bodies. Lough Neagh and Lough Erne in Northern Ireland are eutrophic lakes.

Eutrophication results from the leaching and runoff of nitrate-rich and phosphate-rich artificial fertilisers from agricultural land; and also from the products of the decomposition of organic matter discharged into waterways (see above).

Eutrophication can have the following consequences:

- Increased nitrate and phosphate loading causes massive increases in amounts of algae, particularly blue-green algae (cyanobacteria), i.e. **algal blooms**.
- Dense algal blooms cut down light penetration and use up available ions so that algae die, because they do not receive sufficient light and/or nutrients.
- Dead algae are decomposed by bacteria, which use up oxygen as they respire.
- Oxygen depletion means that many species of invertebrate and fish die.

To prevent further nutrient enrichment, DARD applies the European Nitrates (and Phosphate) Directive to reduce the risk of fertiliser runoff into waterways:

- Inorganic fertilisers can only be applied February–September, during periods of active plant growth, and then not on land bordering a waterway, or on sloping ground or when heavy rain is forecast.
- Soils analysis is recommended, so that the most suitable fertiliser is prepared; fertilisers containing nitrate can only be applied in amounts that match land use; fertilisers containing phosphate can only be applied if soil analysis shows a requirement.

Knowledge check 53

Where discharge of organic matter into waterways is unavoidable (e.g. from septic tanks), reed beds may be constructed at the point of discharge. The common reed has the ability to transfer oxygen from its leaves, down through its stem, and out via its root system. Suggest why reed beds are useful in dealing with minor discharges of organic matter.

Knowledge check 54

Suggest one reason why consequences of eutrophication are:

a more likely to occur in warm water

b less likely to occur in moving water than in still water

Inorganic fertiliser A chemical (artificial or synthetic) fertiliser that releases ions quickly into the soil (compared with organic fertilisers, such as manure, which release nutrients more slowly as they decompose).

Further, phosphates (considered to be the most limiting factor with respect to blue-green algal populations) are removed from sewage works' effluent — all the sewage works around Lough Neagh have this feature.

Habitat and species conservation

Protected sites

Designating special areas for protection is an effective way of conserving the biodiversity of Northern Ireland (see Table 11 and the Department of the Environment website www.doeni.gov.uk).

Table 11 Designations of sites for the protection of habitats and wildlife

Designation	Description
Areas of Special Scientific Interest (ASSIs)	These have been identified by ecological surveys as containing valued plant and animal species. They are managed, using traditional farming methods, with the cooperation of the landowners though legislation (Environmental Order, NI, 2002) is in place that restricts any change in the site that might have a detrimental effect. There are almost 400 ASSIs in Northern Ireland (e.g. Drumlisaleen, Co Fermanagh, consisting of species-rich hay meadows).
Special Areas of Conservation (SACs)	These are given greater protection under European legislation (the Habitats Directive) to help conserve some of the most seriously threatened habitats and species across Europe. There are 57 SACs in Northern Ireland (e.g. Peatlands Park, Co Armagh, containing raised bog and fenland).

Biodiversity Action Plans

The Northern Ireland Biodiversity Action Plan identifies a number of habitat types and species that are rare, vulnerable or declining. Those most at risk have been presented in the NI Priority Habitat and Species Lists (2010). The Priority Habitat List includes 37 distinct types of habitat (e.g. species-rich hedgerows) and for each of these a Habitat Action Plan has been produced. The Priority Species List presently contains 481 species, with over 100 Species Action Plans produced (e.g. for the barn owl).

Action Plans set out the tasks that need to be undertaken to safeguard a habitat or promote the well-being of a species. They also establish the need for various governmental agencies and local bodies to work together to ensure the success of any action. To this end, each local district council has produced its own Local Authority Habitat and Species Action Plan.

Agri-environmental schemes

DARD has established a number of agri-environmental schemes, through a series of legislations and incentives, to conserve habitats and promote species biodiversity:

Knowledge check 55

Explain why organic fertilisers release nutrients over a longer period of time than do inorganic fertilisers.

Knowledge check 56

Why is it important to protect habitats as well as individual species?

Knowledge check 57

What aspects of biodiversity should be considered in any Biodiversity Action Plan?

- Hedge replanting and maintenance to promote hedge diversity (see p. 54).
- Restrictions regarding the application of organic and inorganic fertilisers (see p. 57).
- Leaving ungrazed margins in fields and other farmland areas (field corners and steep slopes) to allow insects and wildflowers to grow, with applications of fertiliser and pesticides prohibited, and so providing a habitat and food source for a range of birds and mammals (see p. 54).
- Retention of winter stubble, from cereals or oilseed rape, to benefit seed-eating birds (such as yellowhammers, tree sparrows and skylarks) that feed on grain left behind after harvesting.
- Planting cereal crops that must not be treated with herbicides ('conservation cereal') to allow a range of weeds in the crop to benefit farmland birds that feed on invertebrates and weed seeds.
- Maintenance of priority habitats on farmland (e.g. species-rich grassland, wetlands) through restrictions — pesticides, fertilisers (organic or inorganic) and drainage are not permitted.
- Other species schemes, for example:
  - leaving uncultivated (fallow) plots to provide lapwings with nesting sites
  - planting a mix of different legumes to supply pollen and nectar for bumblebees and butterflies

> **Exam tip**
>
> You should study a few examples of bird species decline, for example:
>
> - The ground-nesting corncrake disappeared from Northern Ireland as a result of the change from cutting grass towards the end of summer (to make hay) to making several cuts through the year (to make silage, a more productive cattle feed).
> - The planting of winter wheat, which grows early in spring (and so can be harvested earlier) has resulted in a lack of open ground, the preferred nesting situation for skylarks.

Global warming and climate change

Trends in temperature readings from around the world show that **global warming** is taking place. Over the past 130 years, the average global temperature has increased by 0.85°C, with more than half of that increase occurring over the past 35 years (about 0.13°C per decade). The average temperature of the global ocean has increased over the past 50 years (causing seawater to expand and contributing to sea-level rise). There is further supporting evidence of global warming:

- Mountain glaciers have declined, while polar ice melt is accelerating, causing an increase in sea-level rise.
- Arctic permafrost is warming at greater depths.
- Animal and plant species are responding to earlier springs (e.g. earlier bird nesting and earlier flowering).

> **Exam tip**
>
> You are advised to access (via the internet) the Biodiversity Action Plan for your local district council to become aware of the habitats and species discussed in your area.

> **Knowledge check 58**
>
> Explain how the use of fertilisers can lead to a loss of biodiversity.

> **Knowledge check 59**
>
> Permafrost contains a large amount of frozen organic matter. What are the consequences of the permafrost thawing?

Permafrost Subsoil that remains below freezing point throughout the year, occurring chiefly in polar regions. (Thawing of permafrost is an important issue: as permafrost thaws, bacteria become active, decomposing organic matter in the soil and contributing large amounts of CO_2 and methane to the atmosphere — further increasing global warming.)

Throughout the Earth's history the climate has changed, though this is represented by fluctuations over long periods. Now **climate change** is occurring much faster and involves not just increased temperature, but also changes in rainfall, atmospheric pressure and winds. The main effects of climate change are a modification of weather patterns and an increasing number of extreme weather events such as hurricanes, typhoons, floods and droughts.

Future temperatures are expected to rise further. The Intergovernmental Panel on Climate Change (IPCC) in 2014 accepted climate model projections that indicated that during the twenty-first century the global surface temperature is likely to rise a further 1.5°C, at least. Future climate change will differ from region to region. In Northern Ireland, it is projected that by 2050 summers will be hotter and drier, and winters warmer and wetter.

The IPCC (2014) also reported that scientists are 95% certain that most of global warming is caused by increasing concentrations of **greenhouse gases** as a result of human activity. Certainly, carbon dioxide levels have increased — from 300 parts per million (ppm) 200 years ago to the present level of 400 ppm — mainly due to the combustion of fossil fuels, though deforestation is also implicated (less CO_2 taken up in photosynthesis). While CO_2 is regarded as most responsible for an **enhanced greenhouse effect**, there have been increases in other greenhouse gases: methane (CH_4) from livestock and decomposition of faecal and other organic matter; and nitrous oxide (N_2O) from the action of (denitrifying) bacteria on nitrate fertilisers and from motor vehicle emissions. Greenhouse gases act as follows:

- Energy reaching the Earth's surface is radiated back into space at longer wavelengths than the energy arriving from the Sun.
- Some of this longer-wavelength energy (heat) is absorbed by greenhouse gases in the atmosphere, and re-radiated back towards the Earth's surface, with subsequent warming.

Climate is a major factor in determining the distribution of species and the type of ecosystem that exists in an area. So any change in climate may be expected to have profound consequences:

- The distribution of species will shift northwards or to higher altitudes, if the habitat there is suitable: 34 of the 37 British species of dragonfly have expanded their range northwards; the hoopoe, a southern European bird, has spread into northern France and is expected to reach Britain soon.
- The extinction of some species is expected if they are not adapted to the changing environment and have nowhere suitable to move. For example, the crossbill, a bird confined to the very north of Scotland, is a candidate for extinction.
- The polar bear is threatened as the Arctic ice sheet decreases — it is predicted that there may be no summer ice at all.
- Drier conditions will reduce wetlands, important ecosystems for many plant and animal species (e.g. the lapwing, already a priority species).
- Sea level rises will affect many ecosystems and species dependent on the shoreline: salt marsh ecosystems could be lost; turtles could lose the beaches on which they nest.
- The acidification of the oceans (CO_2 is acidic and very soluble in water) is a threat to the coral reef ecosystems.

Knowledge check 60

Temperature rises in the UK could create an environment suitable for mosquitoes previously confined to Africa. What health problems would be associated with an invasion of these mosquitoes?

Knowledge check 61

Explain the rise in carbon dioxide levels in the atmosphere.

Exam tip

Methane is almost 30 times more potent as a greenhouse gas than CO_2, though as its concentration is much lower, and it is shorter lasting in the atmosphere than CO_2, its overall effect is less.

Overall, it is expected that climate change will result in the extinction of many species and the reduced diversity of ecosystems. A major reduction in the use of fossil fuels and a switch to alternative power sources such as wind and nuclear power, allied to increased energy saving schemes, are necessary to avoid major environmental change.

Summary

- Biodiversity is encouraged by: crop rotation (or polyculture); hedge restoration and sensitive management; uncultivated field margins (predator strips); reduced pesticide use within an integrated pest management scheme.
- Release of organic material, such as sewage or farmyard manure, into rivers increases the biological oxygen demand, and reduces the number of species that can inhabit the river. The degree of pollution can be measured by the presence of certain indicator species.
- Runoff and leaching of inorganic fertilisers into waterways causes nutrient enrichment — eutrophication. This increases the growth of algae; their subsequent death and decomposition by aerobic bacteria causes a fall in oxygen levels and many aquatic organisms die.
- Restrictions in the spreading of organic waste and in the application of inorganic fertilisers reduce the risk of runoff. Following soil analysis, fertilisers are prepared to match the crop cultivated.
- Areas of Special Scientific Interest (ASSIs) and Special Areas of Conservation (SACs) conserve important habitats. Habitat and Species Action Plans, and the Department of Agriculture and Rural Development (DARD) agri-environmental schemes, encourage biodiversity.
- Global warming and climate change have resulted from the release of greenhouse gases through human activity. This has brought about changes in the breeding seasons and habitat ranges of many plants and animals.

Questions & Answers

The examination

The AS Unit 2 examination constitutes 37.5% of the AS award; and, since AS represents 40% of A level, contributes 15% to the final A-level outcome. The paper lasts 1 hour 30 minutes and is worth 75 marks. In Section A (60 marks) all the questions are structured, though with a variety of styles. In Section B (15 marks) there is a single question, possibly with several parts, to be answered in continuous prose.

Examiners construct papers to test different assessment objectives (AOs). In the AS Unit 2 paper the approximate marks allocated for each AO are:

- AO1 Knowledge and understanding   27 marks
- AO2 Application of knowledge and understanding   31 marks
- AO3 Analysis, interpretation and evaluation of scientific information, ideas and evidence   17 marks

Skills assessed in questions

The questions in this section test the different assessment objectives. While some assess straightforward knowledge and understanding (AO1), don't be surprised to find something novel — you are being asked to *apply* your understanding (AO2). Some questions will ask you to evaluate experimental and investigative work (AO3).

Since mathematical skills are an important element of biology, the questions include a variety of calculations and graphical work.

Quality of written communication, including accurate use of scientific terms, is assessed in Section B.

About this section

This section consists of questions covering the range of topics covered in the Content Guidance. Following each question there are answers provided by two students of differing ability. Student A consistently performs at grade A/B standard, allowing you to see what high-grade answers look like. Student B makes a lot of mistakes — ones that examiners often encounter — and grades vary between C/D and E/U.

Each question is followed by a brief analysis of what to look out for when answering the question (shown by the icon ⓔ). All student responses are then followed by comments (shown by the icon ⓔ). They provide the correct answers and indicate where difficulties for the student occurred, including lack of detail, lack of clarity, misconceptions, irrelevance, poor reading of questions and mistaken meanings of examination terms. The comments suggest areas for improvement.

In using this section try the questions before looking at the students' responses or the comments, which you can then use to mark your work. Check where your own answers might have been improved.

The CCEA biology specification is available from www.ccea.org.uk. Apart from the *subject content*, you must familiarise yourself with the *mathematical skills* shown in section 4.7 and the *command terms* used in examinations (such as *explain*, *describe* and *suggest*) shown in Appendix 1. The website will also allow you to access *past papers* and *mark schemes*.

■ Section A Structured questions

Principles of exchange and transport

Question 1 Body size and shape

(a) (i) Large organisms have a greater surface area than small organisms, yet their body surface is less able to supply their need for oxygen. Explain why it is the surface area-to-volume ratio that is important in determining whether an organism satisfies its oxygen needs. (3 marks)

(ii) Explain why terrestrial animals, such as mammals, would have a problem absorbing oxygen through the body surface. (1 mark)

(b) (i) Define the term diffusion. (1 mark)

(ii) Explain why molecules are moved much faster by mass flow than by diffusion. (2 marks)

(iii) Explain why flatworms (thickness 0.3 mm) lack a blood transport system, while earthworms (thickness 5 mm) possess a blood system. (2 marks)

Total: 9 marks

ⓔ This is mostly knowledge and understanding, though the last part requires you to apply your understanding. You will need to be precise in your wording — students frequently give vague answers on this topic. Also, make sure that you distinguish between the need for a specialised absorptive surface and the need for a transport system.

Student A

(a) (i) The volume determines the amount of O_2 required, while the surface area determines the amount of O_2 absorbed. ✓ The SA:V ratio needs to be large enough for sufficient O_2 to be supplied to meet the organism's needs. ✓ Smaller organisms have a much larger SA:V ratio than large organisms. ✓

(ii) Terrestrial animals are waterproof to prevent dehydration. Gases cannot pass through a waterproof surface. ✓ [a]

(b) (i) Diffusion is the net movement of molecules from a region of high concentration to a region of lower concentration. ✓

(ii) Diffusion relies on the random movement of molecules. In mass flow the molecules are all moved in the same direction ✓ since they are driven by some type of pump ✓.

(iii) The thin shape of the flatworm means that all cells are close enough to the body surface for sufficient oxygen to reach them by diffusion. ✓ In an earthworm the innermost tissues would not get sufficient oxygen if relying on diffusion so it needs a blood system to speed up the movement of molecules. ✓ [a]

ⓔ 9/9 marks awarded ⓐ All answers correct, for full marks.

ⓔ 3/9 marks awarded ⓐ The link between SA:V ratio and size is correct for 1 mark, but just saying that SA:V ratio determines efficiency does not explain why. ⓑ Student B presumably understands that terrestrial animals have a waterproof surface so 1 mark awarded. ⓒ No mark awarded because it is important to say that diffusion is a *net* movement. ⓓ Student B understands that mass flow involves some sort of pump, but has not included anything about the direction of movement. ⓔ This is true, but not relevant regarding the need for a transport system. No mark awarded.

Gas exchange in plants and mammals

Question 2 Gas exchange in plants

(a) Fick's law states that:

$$\text{rate of diffusion} \propto \frac{\text{surface area} \times \text{concentration gradient}}{\text{diffusion distance}}$$

Explain how a leaf is adapted to comply with Fick's law in maximising gas exchange for photosynthesis. (3 marks)

(b) (i) The amount of oxygen released by a plant is a measure of the net rate of photosynthesis. Explain what is meant by the 'net rate of photosynthesis'. (1 mark)

(ii) Explain what is meant by the 'compensation point'. (1 mark)

(c) Hydrophytes are flowering plants adapted for life in water.

(i) Describe and explain two adaptations of hydrophytes. (2 marks)

(ii) Few flowering plants can survive in the marine environment. Suggest why. (1 mark)

Total: 8 marks

ⓔ In part (a), be careful that you apply your understanding about Fick's law to gas exchange in the leaf. Part (b) is testing your understanding of biological terms. In part (c) (i), do not just list two adaptations — you must also explain how they operate as adaptations; in part (c) (ii), you need to think about conditions in a marine environment.

Student A

(a) Leaves are flattened to provide a large surface area. ✓ As carbon dioxide is used by mesophyll cells there is a low concentration within the leaf, so stomata are open to ensure a fresh supply and maintain a concentration gradient. ✓ Leaves are thin so the diffusion distance is small. ✓ a

(b) (i) The net rate of photosynthesis is the difference between the rate of photosynthesis and the rate of respiration. ✓

(ii) The compensation point is where the rate of photosynthesis equals the rate of respiration. ✓ a

(c) (i) Hydrophytes have aerenchyma through which oxygen diffuses to the roots for respiration. ✓ The leaves may float and have stomata on the upper surface so that carbon dioxide is supplied for the leaf's photosynthesis and oxygen for the whole plant's respiration. ✓ a

(ii) Flowering plants cannot withstand the salty conditions. ✗ b

ⓔ **7/8 marks awarded** a All correct for a total of 7 marks. b Student A should have tried to explain that the salty environment would have caused dehydration through loss of water by osmosis.

Student B

(a) Leaves are thin and flat. Being thin decreases the diffusion distance ✓, while being flattened provides a large surface area ✓. They also have stomata. a

(b) (i) Net photosynthesis equals photosynthesis less respiration. ✓ b

(ii) At the compensation point, they are equal. ✗ c

(c) (i) Hydrophytes have stomata on the upper surface of leaves to allow gas exchange between the air and the mesophyll. ✓ d

(ii) In salt water flowering plants would lose water by osmosis. ✓ e

ⓔ **5/8 marks awarded** a Student B has explained how the leaf provides a large surface area and a small diffusion distance. However, there is no attempt to explain how stomata maintain a concentration gradient. b Correct, for 1 mark. c The examiner cannot assume that 'they' refers to photosynthesis and respiration. No mark awarded. d Only one adaptation has been provided, so only 1 mark. e Student B has been able to transfer understanding from AS1, so 1 mark gained.

Question 3 Gas exchange in mammals

(a) The diagram below shows a section through an alveolus and associated structures.

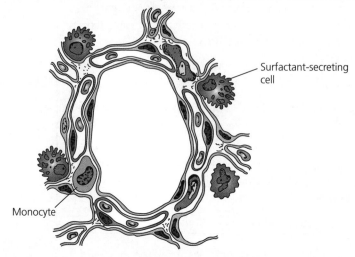

(i) State one adaptation, shown in the diagram, that facilitates gaseous exchange. Explain how this adaptation aids gaseous exchange. (2 marks)

(ii) State one adaptation, *not* shown in the diagram, which is needed for efficient gaseous exchange. Explain how this adaptation aids gaseous exchange. (2 marks)

(iii) State the role of each of the following:
   ▪ surfactant
   ▪ the monocyte (2 marks)

(b) Use the following information to calculate the total surface area of the alveoli in the lungs of an individual.

Each lung has 350 million alveoli. The average diameter of each alveolus is 0.2 mm. The alveoli are not complete spheres since each has an opening that reduces the surface area by 20%. The formula for calculating the surface area of a sphere is $4\pi r^2$, where $\pi = 3.14$. (3 marks)

Total: 9 marks

ⓔ Part (a), (i) and (ii), might seem straightforward, but you need to be careful because (i) asks for an adaptation *shown* in the diagram while (ii) asks about one *not shown*. Part (a) (iii) asks about the role of features within the alveolus that are not involved directly in gaseous exchange, and tests your understanding of biological terms. Part (b) presents a numeric problem, which involves a number of steps, so read the information carefully.

Student A

(a) (i) The wall of the alveolus consists of a squamous epithelium in which the cells are thin. ✓ a

(ii) A layer of moisture lines each alveolus. ✓ Gases must dissolve in water before they can diffuse through the cells. ✓ b

(iii) Surfactant reduces the surface tension preventing the polar water molecules on the inner walls of the alveoli from sticking together. ✓ The monocyte moves into the alveolus where it becomes a macrophage that engulfs any bacteria. ✓ c

(b) SA of each alveolus = $4 \times 3.14 \times 0.1 \times 0.1 \times 0.8 \, mm^2 = 0.10048 \, mm^2$ ✓

SA of each lung = $350 \times 10^6 \times 0.10048 \, mm^2 = 35\,168\,000 \, mm^2$ ✓

SA of two lungs = $70\,336\,000 \, mm^2 = 70.3 \, m^2$ ✓ d

ⓔ 8/9 marks awarded a Thinness is correct for 1 mark, but Student A should have said that this reduces the diffusion distance. b A complete answer, for 2 marks. c Excellent answers, for 2 marks. d This is well set out and so it is easy to follow the steps. Student A has used a factor of 0.8 (80%) to account for the alveolar openings, and has given the answer in square metres (even though that was not required) and, appropriately, to three significant figures. 3 marks gained.

Student B

(a) (i) Thin membranes. ✗ This results in more diffusion. ✗ a

(ii) There are millions of alveoli making up the lungs ✓ and these greatly increase the surface area over which gas exchange can take place ✓. b

(iii) Surfactant — to reduce the surface tension. ✓ c
Monocyte — to digest any worn out organelles. ✗ d

(b) $350 \times 1000000 \times 4 \times 3.14 \times 0.2 \times 0.2 \times 2 \, mm^2 = 351\,680\,000 \, mm^2$ ✗

Less 20% for openings = $70\,336\,000 \, mm^2$ ✓

So total surface area of lungs = $281\,344\,000 \, mm^2$ ✓ e

ⓔ 5/9 marks awarded a The term 'membrane' can be confusing. Student B should have noted that the cells of the alveolar walls are thin — they form a squamous epithelium. It is not correct to say that there is more diffusion — the distance over which gases have to diffuse is reduced. Student B fails to score. b The diagram does not show the vast number of alveoli, so this is correct, as is the explanation. 2 marks awarded. c This is sufficient to obtain the mark, though a fuller answer would note that a reduction in surface tension prevents the walls of the alveoli sticking together and collapsing during exhalation. d It is lysosomes within cells that 'digest worn-out organelles'. Monocytes develop into macrophages that ingest any bacteria present. e Student B has made one operational mistake, using diameter rather than radius in the calculation. This mistake is not further penalised, so 2 marks are gained. While the answer was not required in m^2 and to three significant figures, this might be the case in other exam questions.

Question 4 Breathing in mammals

(a) Explain why it is necessary for mammals to ventilate their lungs.   (1 mark)

(b) Describe and explain the process of exhalation in a mammal.   (5 marks)

(c) Describe the condition emphysema and its effect on the lungs.   (1 mark)

(d) The graph below shows the effect of smoking on lung function.

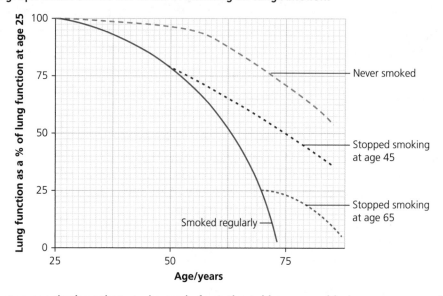

Give two conclusions that can be made from the evidence provided.   (2 marks)

Total: 9 marks

ⓔ Parts (a), (b) and (c) require you to recall information on the functioning of the lungs and is therefore assessing AO1. Notice the mark tariff in part (b) — your answer must be sufficiently detailed, and since you are asked to *describe and explain* you need to make clear links between cause and effect. Part (c) tests your ability to interpret evidence on a graph — make sure you study the graph carefully before answering.

Student A

(a) Ventilation establishes a relatively high concentration of oxygen and low concentration of carbon dioxide within the bronchial tree so that there is a high diffusion gradient with the alveolar air. ✓ ⓐ

(b) The muscle of the diaphragm and the external intercostal muscles relax. ✓ So the diaphragm returns to its normal dome-shaped position and the rib cage drops. ✓ Consequently, the volume of the thoracic cavity decreases ✓, causing the pressure to increase ✓. When the pressure within the lungs exceeds that in the atmosphere, air is forced out. Also, the lungs contain elastic tissue, which, through recoil, forces air out of the alveoli. ✓ ⓐ

(c) In emphysema, the alveolar walls break down, decreasing the surface area for gas exchange and leading to breathlessness. ✓ ⓐ

(d) A person who gives up smoking when 45 has worse lung function than a non-smoker, but much better than those who continue smoking. ✓ For a person who gives up smoking at 65, lung function worsens more slowly than if they continued smoking. ✓

ℯ 9/9 marks awarded a All correct, for full marks.

Student B

(a) Ventilation of the lungs moves fresh air rapidly towards the alveoli, so aiding the maintenance of a high diffusion gradient for both oxygen and carbon dioxide. ✓ a

(b) The intercostal muscles relax, so the rib cage moves down and in. ✓ The diaphragm relaxes and so moves up. ✓ As a result air is forced out of the lungs. b

(c) The alveoli have broken down giving fewer and larger alveoli, which means that there is a lower surface area for gas exchange. ✓ c

(d) People who smoke at any age have lower lung function than people who never smoked. ✓ This is because they develop emphysema. ✗ d

ℯ 5/9 marks awarded a Correct, for 1 mark. b Student B understands the muscles involved and what happens when they relax, achieving 2 marks. However, the effects on the thoracic volume and pressure are ignored, as is the importance of lung elasticity. c Correct, for 1 mark. d The first point is valid and gains 1 mark. However, the second point is not evident from the information in the graph.

Transport and transpiration in plants

Question 5 The structure of a root

The photograph below is of part of a transverse section through the root of a buttercup (*Ranunculus*).

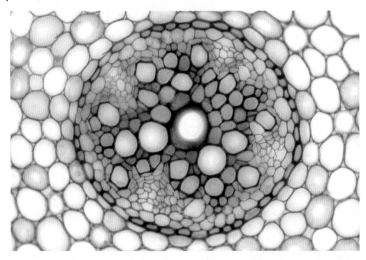

Draw a block diagram to show the tissue layers in the root as shown in the photograph. Label the drawing to identify at least four structures.

(8 marks)

Total: 8 marks

(e) This question assesses drawing skills and the identification of root structures. Study the photomicrograph to identify the obvious tissue layers and then draw them with smooth, continuous and non-sketchy lines. You must draw the structures in the photograph provided, *not* provide a textbook diagram of a root. The proportionality of structures must be reasonably accurate. There are 4 marks for drawing skills and 4 marks for identification.

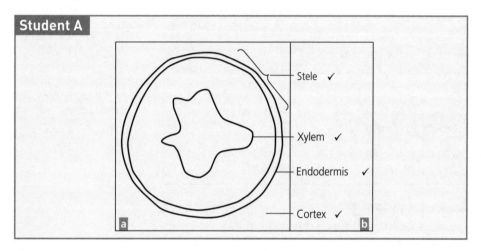

Student A

(e) **6/8 marks awarded** a This is a block diagram showing only the xylem and endodermis tissue layers and so it lacks completeness. ✗ It is an attempt to represent the photograph ✓, but it lacks the proportionality of the features shown ✗. The lines drawn are smooth and continuous, not sketchy. ✓ Student A gains 2 marks for drawing skills. b 4 marks are awarded for correctly labelling stele, xylem, endodermis and cortex, even though the phloem is not drawn and so not recognised.

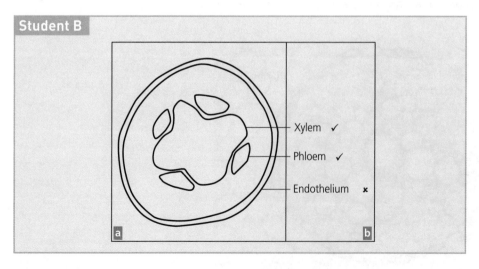

Student B

ⓔ **5/8 marks awarded** ⓐ This is a block diagram of the tissues obvious in the photograph. ✓ It is also an attempt to represent the photograph ✓ though it lacks the proportionality of the features shown ✗ since the xylem region is too small within the stele. The lines drawn are smooth and continuous, and not sketchy. ✓ Student B gains 3 marks for drawing skills. ⓑ The labels for xylem and phloem are correct, for 2 marks. However, endothelium is not acceptable for endodermis. Other recognisable features include the stele (vascular cylinder) and the cortex.

Question 6 Xylem and phloem

(a) Xylem vessel cell walls are thickened by lignin.

    (i) Explain the function of the lignified thickening in xylem vessels. (2 marks)

    (ii) Explain why the xylem vessels in young stems have a pattern of thickening that is annular or spiral. (2 marks)

    (iii) In mature regions of the stem, xylem vessels have a pitted pattern of lignification. Explain the function of the pits. (2 marks)

(b) A series of experiments was carried out to investigate translocation in phloem tissue. In the experiments, a leaf was supplied with radioactive carbon dioxide, incorporating ^{14}C as a tracer. In two of the experiments, the plants were ringed — a ring of surface tissues including phloem was removed. The results of the experiments are shown in the diagram below: the regions ringed are shown by small black bars; numbers on the diagram indicate the amount of radioactive carbon detected in each part of the plant.

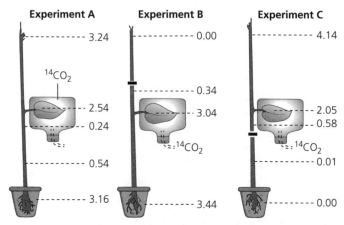

    (i) Name the molecule into which the radioactive ^{14}C will be incorporated within the phloem. (1 mark)

    (ii) Explain the purpose of experiment A in the diagram. (1 mark)

    (iii) What conclusions can be drawn from the experimental results? (3 marks)

    (iv) Similar experiments show that labelled molecules in the phloem move at a rate of $35\,mm\,min^{-1}$, a rate much higher than can be explained by diffusion. What does this suggest about the process of translocation? (1 mark)

Total: 12 marks

Questions & Answers

(e) Part (a) is testing your understanding (AO1) of xylem vessels — just make sure that you answer the questions fully. Part (b) presents an investigation (AO3) of phloem translocation. You will need to study the diagram carefully and ensure that your answers are *relevant to the questions asked*. Notice that the mark value for (b) (iii) is 3 so there must be at least three valid points.

Student A

(a) (i) The lignin thickening prevents the vessels collapsing ✓ under the negative pressure created by transpiration ✓.

(ii) Annular or spiral thickening allows the vessels to elongate ✓ in the region of growth within the young stem ✓.

(iii) The pits are pores that allow water to enter and leave the xylem vessels ✓, since lignin is impermeable to water ✓. [a]

(b) (i) Sucrose. ✓ [b]

(ii) This is a control experiment. ✗ [c]

(iii) Sucrose is transported in phloem, both up and down. ✓ Where possible, sucrose accumulates in the growing plant tip and in the roots. ✓ Ringing prevents movement of sucrose since the phloem is removed. ✓ [d]

(iv) Sucrose is pumped into the sieve tubes from the companion cells. ✗ [e]

(e) **10/12 marks awarded** [a] All answers complete, for a total of 6 marks. [b] Correct, for 1 mark. [c] Student A needed to explain how it acts as a control. No mark awarded. [d] Three valid points, for 3 marks. [e] This is correct but is not an answer to the question asked. No mark gained.

Student B

(a) (i) The lignin gives strength and prevents collapsing. ✓ [a]

(ii) Rings and spirals of lignin allow the vessels to stretch. ✓ [b]

(iii) The lignin layer is waterproof ✓, so pits are needed to allow water in and out ✓. [c]

(b) (i) Glucose. ✗ [d]

(ii) This is a comparative experiment to show where sucrose is translocated in an intact plant. ✓ [e]

(iii) Translocation occurs both upwards and downwards. ✓ [f]

(iv) Translocation is an active process involving energy expenditure. ✓ [g]

(e) **7/12 marks awarded** [a] Student B has not explained why collapsing would need to be prevented, so only 1 mark is awarded. [b] There is no explanation of why vessels would need to stretch, so only 1 mark scored. [c] A full answer, for 2 marks. [d] Carbohydrate may occur as glucose in the leaf but is translocated as sucrose.

No mark gained. e A complete answer, for 1 mark. i Only one point is provided so only 1 mark can be scored. Student B would appear to have ignored the mark allocation. g Correct, for 1 mark.

Question 7 Factors affecting transpiration

The graphs below show the results of an investigation into the effects of two factors — wind speed and relative humidity — on the rate of transpiration.

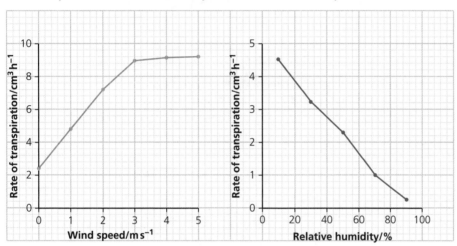

(a) (i) Describe the effects of wind speed on the rate of transpiration. (2 marks)

(ii) Explain these effects. (2 marks)

(iii) Explain why the rate of transpiration is not zero when the wind speed is zero. (1 mark)

(b) Describe and explain the effect of relative humidity on the rate of transpiration. (2 marks)

(c) In both experiments the plant was fully illuminated. Explain what would have happened if the experiments had been carried out in the dark. (2 marks)

(d) Name one external factor, other than relative humidity and light intensity, which should have been kept constant during the experiment on the effect of wind speed on the rate of transpiration. (1 mark)

Total: 10 marks

e This question requires you to interpret an investigation of factors influencing transpiration (testing a combination of AO2 and AO3). Remember that when asked to *describe*, you should present in words any trends shown, whereas if asked to *explain*, you must give reasons for what is happening.

Student A

(a) (i) As wind speed increases to $3\,m^2\,s^{-1}$ the rate of transpiration increases. ✓ Above $3\,m^2\,s^{-1}$ a further increase in wind speed has no further effect on the rate of transpiration. ✓

(ii) Wind blows away the humid diffusion shells outside the stomata, so increasing the diffusion gradient for water vapour to diffuse out of the leaves. ✓ Above a wind speed of $3\,m^2\,s^{-1}$ the diffusion shells have been fully removed so that the diffusion gradient cannot be further increased. ✓

(iii) Even if there is no wind, water vapour will still diffuse out of the leaves. ✓ ⓐ

(b) The higher the relative humidity the lower the rate of diffusion. ✓ This is because if the external humidity is high then diffusion out of the leaves is reduced. ✓ ⓐ

(c) In the dark the stomata would close. ✓ ⓑ

(d) The number of stomata. ✗ ⓒ

ⓔ **8/10 marks awarded** ⓐ All answers complete, for a total of 7 marks. ⓑ Student A has omitted to say anything about the effect on the experimental results. Only 1 mark awarded. ⓒ The question asks for an external factor; this is an internal factor. No mark gained.

Student B

(a) (i) Transpiration increases as the wind speed increases. ✓ ⓐ

(ii) The wind blows away the diffusion shells outside the stomata. ✓ ⓑ

(iii) Water would still diffuse out of the stomata if there was no wind. ✓ ⓒ

(b) The more humid the conditions the lower the rate of diffusion. ✓ ⓓ

(c) In the dark the stomata close ✓ and, since this is the main route of water loss, transpiration is greatly reduced ✓. ⓔ

(d) A fan is used to control wind speed. ✗ ⓕ

ⓔ **6/10 marks awarded** ⓐ Student B has failed to describe the plateau part of the graph, so only 1 mark is awarded. ⓑ Again, the plateau part of the graph has been ignored, so only 1 mark gained. ⓒ Correct, for 1 mark. ⓓ This is a correct description of the effect of humidity but there is no attempt to explain the effect. 1 mark awarded. ⓔ A full answer, for 2 marks. ⓕ Student B has not read the question properly and has said how wind speed might be controlled rather than identified another factor to control. No mark awarded. Overall, Student B failed to provide sufficiently detailed answers in this question.

Circulatory system in mammals

Question 8 The cardiac cycle

The graph below shows pressure changes that take place in the left side of the heart during one complete cardiac cycle.

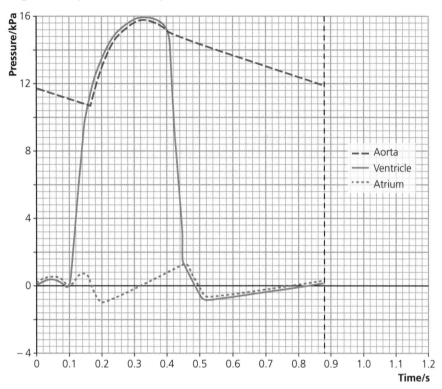

(a) Determine the time at which each of the following valves close, in each case, giving the reason for your answer:

   (i) atrioventricular (bicuspid) valve   (2 marks)

   (ii) aortic valve   (2 marks)

(b) Explain how the structure of each valve type allows its closure.   (2 marks)

(c) Explain the changes in the atrial pressure from

   ■ 0–0.05 seconds

   ■ 0.2–0.45 seconds   (2 marks)

(d) Explain the increase in the aortic pressure between 0.16 and 0.3 seconds.   (2 marks)

(e) The maximum pressure in the left ventricle is 16 kPa, while the pressure in the right ventricle reaches a maximum of only 3.5 kPa. Explain this difference.   (2 marks)

(f) Excitation waves spread through the heart during each cycle of contraction.

   (i) Name the specialised tissue through which the wave of excitation passes up the ventricular muscle.   (1 mark)

   (ii) How can irregularities in the excitation of the heart be medically assessed?   (1 mark)

                                                       Total: 14 marks

@ Study the pressure changes carefully before you attempt to answer the questions. To answer (a) correctly, you must *work out* what changes in blood pressure cause the valves to close. Part (b) tests your understanding of valve structure within the heart. To answer (c) and (d) well, you must apply your understanding of the cardiac cycle to *explain* the changes in pressure. Parts (e) and (f) should be relatively straightforward.

Student A

(a) (i) The atrioventricular valve closes at 0.1 s ✓, as contraction of the ventricle causes its pressure to increase above that in the atrium. ✓

(ii) The aortic valve closes at 0.4 s ✓, as the ventricle relaxes and the ventricular pressure decreases and becomes less than that in the aorta. ✓ a

(b) The atrioventricular valves consist of flaps, which are forced towards the atria but prevented from turning inside out by the chordae tendinae. ✓ The aortic valve consists of pockets, which fill with blood when the pressure in the aorta is higher than in the ventricle. ✓ a

(c) From 0 to 0.05 s, the atria are contracting. ✓ From 0.2 s, the pressure in the atrium increases gradually because the atrium is passively filling with blood from the pulmonary vein. ✓ a

(d) Since the ventricular pressure is greater than the aortic pressure ✓ the semilunar valve is forced open ✓ and blood is pumped into the aorta, increasing pressure there. a

(e) The muscular wall of the left ventricle is much thicker, to pump blood into the aorta. ✓ b

(f) (i) Purkinje fibres ✓

(ii) Electrocardiography ✓ c

@ **13/14 marks awarded** a These parts are well answered, for a total of 10 marks. b This is correct for 1 mark, but no reference is made to the *reason* for the higher pressure generated — see Student B's response and the comment. c Both parts correct, for 2 marks.

Student B

(a) (i) 0.1 s. ✓ This is when the atrium contracts, forcing blood into the ventricle and therefore increasing its pressure. ✗ a

(ii) 0.4 s. ✓ This is when the aortic pressure is at its highest, so the valve closes to prevent backflow. ✗ b

(b) The AV valve closes when the pressure in the ventricle is greater than in the atrium. ✗ The aortic valve closes when the pressure in the aorta is greater than in the ventricle. ✗ c

(c) Between 0.0 s and 0.05 s blood is returning to the atrium. ✗ Between 0.2 s and 0.45 s the atrium is contracting and pumping blood into the ventricle. ✗ d

(d) Pressure increases as blood is forced into the aorta from the ventricle, where the pressure is higher. ✓ e

(e) This is because the wall of the left ventricle contains more muscle. ✓ This is required to produce a higher pressure to pump the blood around the whole body. ✓ f

(f) **(i)** Purkyne fibres ✓ g

(ii) Angiography ✗ h

ⓔ **6/14 marks awarded** a The time is correct, for 1 mark. However, the reason given is not correct. At this time, the pressure in the ventricle increases to become greater than that in the atrium, so forcing the atrioventricular valve closed. b Again, the time is correct, for 1 mark, but the reason given is not. Most particularly, a valve does *not* close to prevent backflow. In this case, it closes because the pressure in the ventricle drops below that in the aorta. As a result, backflow is prevented. c Student B is confused — the question asks about the structure of the valves, not the reasons for valve closure; indeed, these are the answers to part (a). No marks awarded. d More confusion — the answers are the wrong way round! No marks scored. e This is correct, for 1 mark, though a more complete answer would have noted that the semilunar valve at the base of the aorta has been forced open. f Both points are correct. An alternative answer would be to note that a low pulmonary pressure is required because slow blood flow increases the efficiency of gas exchange in the lungs, and also prevents fluid being forced into the alveoli. Student B scores 2 marks. g Correct — Purkinje is also known as Purkyne — for 1 mark. h Student B has confused angiography with electrocardiography — this is simple learning. No mark.

Question 9 Blood vessels

(a) The diagram below shows an artery and vein.

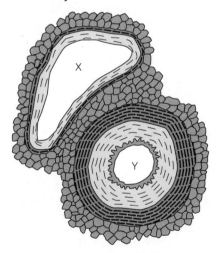

(i) Identify which of the two, X or Y, represents an artery. (1 mark)

(ii) Describe two features of an artery and explain how each is adaptive in the functioning of the artery. (2 marks)

(iii) Describe one feature of a vein and explain how it is adaptive in the functioning of the vein. (1 mark)

(b) Outline the sequence of events that occur during the process of atherosclerosis. (3 marks)

(c) Describe how an angiogram may be used to diagnose damage to arteries. (2 marks)

Total: 9 marks

ⓔ In part (a), you are presented with a diagram that you are expected to interpret. This is relatively straightforward, although in (ii) and (iii) your answers must *explain* how features are adaptive, i.e. give reasons as to how they are adaptive. Part (b) is simple recall of the process of atherosclerosis — a full, well-sequenced answer is required. In part (c), you need to remember how an angiogram is used medically to identify injury to arteries.

Student A

(a) (i) X ✗ ⓐ

(ii) Elastic fibres to allow for expansion and recoil. ✓ Possesses a small lumen. ✗ ⓑ

(iii) It has a large lumen, which ensures that there is little resistance to blood entering and passing through the vein. ✓ ⓒ

(b) Atherosclerosis starts with injury to the endothelium of an artery ✓, because of high blood pressure or toxins in cigarette smoke. Monocytes invade the artery wall and develop into macrophages, which accumulate fats and cholesterol, forming an atheroma. ✓ Fibrous tissue and calcium salts add themselves to the atheroma, creating a hard plaque, which narrows the lumen of the artery. ✓ ⓒ

(c) Atherosclerosis is recognised in an angiogram as the damaged artery would be narrowed ✓ through the build-up of plaque. An aneurysm would be recognised as the artery would be balloon-like. ✓ ⓒ

ⓔ **7/9 marks awarded** ⓐ This is incorrect, although from the answers below this seems like a slip. ⓑ There is no attempt to explain why having a small lumen is an adaptive feature — it ensures that high blood pressure is maintained. Student A scores only 1 mark. ⓒ These parts are well answered, for 6 marks.

Student B

(a) (i) Y ✓ ⓐ

(ii) Elastic fibres to withstand high pressure and allow recoil. ✓ Collagen fibres for strength. ✗ ⓑ

(iii) Possesses pocket valves. ✗ ⓒ

(b) Atherosclerosis occurs when there is damage to an artery wall and fats and cholesterol form deposits in the wall behind the site of injury. ✓ This causes the wall to be less elastic. ⓓ

(c) Atherosclerosis can be diagnosed because the blood vessel will be narrowed. ✓ An aneurysm is diagnosed as a total block due to the presence of a blood clot. ✗ ⓔ

ⓔ 4/9 marks awarded **ⓐ** Correct, for 1 mark. The artery has a thicker wall and a smaller lumen than a vein. **ⓑ** The first adaptation is correct. However, the reference to collagen is not a specific adaptation in the functioning of the artery. The artery has a thick layer of smooth muscle which, when it contracts, reduces the blood flow to the organ that it supplies. Student B scores 1 mark. **ⓒ** This is correct, but Student B makes no attempt to explain why valves are adaptive — for example, that they ensure one-way flow towards the heart — and fails to score. **ⓓ** This lacks detail — there is no mention of the initial damage being to the lining of the artery wall; the formation of a plaque is ignored. Basically, Student B has described the formation of an atheroma. 1 mark scored. **ⓔ** The first answer is correct but Student B has confused aneurysm with thrombosis. 1 mark awarded.

Question 10 Blood cells and blood clotting

(a) The diagram below shows three types of white blood cell, A, B and C. Identify each type and state its function. (3 marks)

A   B   C

(b) The diagram below summarises the process of clotting. Identify the components of each step. (2 marks)

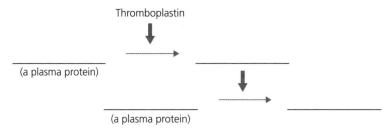

Thromboplastin

_____ (a plasma protein)

_____ (a plasma protein)

Total: 5 marks

ⓔ This should be straightforward recall and is the type of question that you might expect early in an AS Unit 2 paper, targeted at lower-grade candidates. Notice that in part (a) two answers are required for each mark — you must both identify and give the function of each cell type. In part (b), you must provide the names of both components of each step.

Student A

(a) A = lymphocyte — produces antibodies or destroys foreign cells; ✓
B = monocyte — develops into a macrophage, which is phagocytic; ✓
C = polymorph, which is phagocytic. ✓ **ⓐ**

(b) prothrombin → thrombin; ✓ then fibrinogen → fibrin ✓ **ⓐ**

ⓔ 5/5 marks awarded **ⓐ** All correct, for full marks.

ⓔ **2/5 marks awarded** ⓐ Student B has confused the lymphocyte with the monocyte — both possess prominent nuclei, but the monocyte's is kidney shaped. Further, Student B has failed to give the functions of these cells. The answer to C is correct, for 1 mark. ⓑ The first step is correct for the mark. Fibrinogen converts to fibrin, which forms the fibres of the clot. However, the technical term is required.

Question 11 Oxygen dissociation curves

The graph below shows oxygen dissociation curves for haemoglobin at two different partial pressures of carbon dioxide (pCO_2), and for myoglobin. The partial pressures of oxygen (pO_2) in the lungs, in muscle during moderate exercise, and in muscle during strenuous exercise, are indicated.

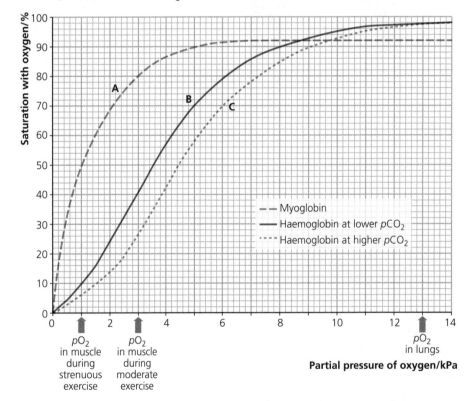

(a) With reference to both pO_2 and pCO_2 levels, explain why haemoglobin unloads oxygen to actively respiring tissues, such as exercising muscle. (3 marks)

(b) Describe and explain the significance of the difference in affinity for oxygen between haemoglobin and myoglobin. (3 marks)

(c) The haemoglobin dissociation curve for a llama, a mammal found at high altitudes, lies to the left of human haemoglobin. Explain the advantage of this to the llama. (3 marks)

(d) If curve B represents the dissociation curve for haemoglobin at pH 7.4, would curve C represent the dissociation curve at a higher or a lower pH? Explain your answer. (1 mark)

Total: 10 marks

ⓔ This question tests understanding of the topic, rather than just recall. In parts (a) and (b), you must use the information in the graph and your own understanding of dissociation curves — detailed answers are required since each part is worth 3 marks. In part (c), think first of all about oxygen levels at high altitude — again, a full answer is required. Part (d) requires you to think about the link between pCO_2 and pH.

Student A

(a) In respiring tissues, oxygen is used up and pO_2 is reduced to 3 kPa. During strenuous exercise, this can become as low as 1 kPa. ✓ At lower pO_2 haemoglobin cannot hold on to its oxygen and so it is unloaded. ✓ Respiration produces carbon dioxide and so pCO_2 increases, with the effect that haemoglobin releases even more oxygen. ✓ ⓐ

(b) Myoglobin has a much higher affinity for oxygen than haemoglobin ✓ and only releases its oxygen within muscle when the pO_2 is very low ✓, such as during strenuous exercise, when oxygen demand is high. ✓ ⓐ

(c) At high altitude, the habitat of the llama, atmospheric pO_2 is lower than at low altitude. ✓ An oxygen dissociation curve that lies to the left indicates a higher affinity for oxygen. ✓ So the llama's haemoglobin is adapted to saturate with oxygen at lower pO_2 levels. ✓ ⓐ

(d) Curve C would be at a lower pH, since CO_2 is an acidic gas. ✓ ⓐ

ⓔ **10/10 marks awarded** ⓐ Excellent, detailed answers throughout.

Student B

(a) In the tissues, oxygen is used up and in such circumstances haemoglobin unloads its oxygen. ✓ ⓐ

(b) Myoglobin has a higher affinity for oxygen. ✓ It enables aerobic respiration to continue for longer. ✗ ⓑ

(c) A haemoglobin curve to the left means that it has a greater affinity for oxygen. ✓ This means that it takes up oxygen more easily. ✓ ⓒ

(d) At a lower pH. ✗ ⓓ

ⓔ **4/10 marks awarded** ⓐ This is fine for 1 mark, but does not go far enough since there is no mention of respiration or the effects of an increase in pCO_2 on oxygen unloading. For 3 marks a fuller answer is necessary. ⓑ The first part describes the difference and is awarded 1 mark. However, the second part, while true, does not *explain* the situation. **c** Two valid points for 2 marks, but Student B has failed to mention the significant point that pO_2 is low at high altitude. **d** This is correct, but no reasoning has been presented and so the answer fails to score.

Adaptations of organisms

Question 12 Adaptations of shags and cormorants

The shag and the cormorant — illustrated below — are two common and rather similar coastal birds.

Shag (*Phalacrocorax carbo*)   **Cormorant** (*P. aristotelis*)

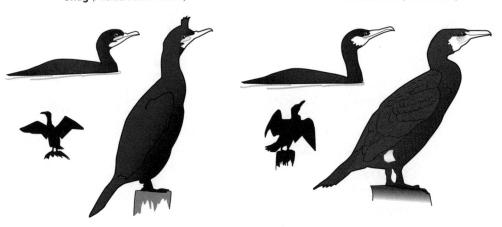

Both eat fish, which they catch by diving from the surface. The table below shows the types and locations of their prey.

Prey		% of prey taken by	
		shag	cormorant
Surface-swimming prey	Sand-eels	33	0
	Herring	49	1
Bottom-feeding prey	Flatfish	1	26
	Shrimps, prawns	2	33

(a) Describe the ecological niches of the shag and the cormorant. Why is it important that they have different niches? (3 marks)

The cormorant has several structural adaptations. For example, for swimming underwater it has large and fully webbed feet.

(b) What further adaptation (visible in the illustrations above) does the cormorant have for catching fish? (1 mark)

Cormorants have two layers of feathers: the inner body plumage is oiled and remains dry near the skin; the outer feathers are designed to get thoroughly wet and become waterlogged when the bird submerges to hunt.

(c) Suggest an advantage for the outer feathers becoming waterlogged during diving. (1 mark)

After spending time in the water, cormorants exhibit wing-spreading behaviour, stretching out their wings to dry their feathers.

(d) Suggest an explanation for this 'wing-spreading behaviour'. (1 mark)

While physiological adaptations for diving have been little studied in cormorants, it is known that they can dive down to 50 metres and that dives can last for several minutes.

(e) Suggest one physiological feature that would be appropriate to study. Explain your answer. (2 marks)

Total: 8 marks

e In part (a), you are asked to use the data provided to describe the 'ecological niches' of two species. Notice that the last three parts use the command term *suggest*. This means that you need to give plausible explanations for something that is unfamiliar. This is AO2 — think about the information provided and don't just try to remember something that you have learned.

Student A

(a) The shag is a coastal bird, feeding mainly on fish in the surface water. ✓ The cormorant is similar but feeds mostly on fish on the sea-bed. ✓ They therefore have different niches and so they avoid competing with each other. ✓ a

(b) They have long, hooked bills for grasping their prey. ✓ a

(c) If the outer feathers absorb water the cormorant will be able to dive faster to chase fish at the sea-bed. ✓ Otherwise the cormorant would be too buoyant. a

(d) This could be thermoregulatory. Allowing the Sun to dry feathers means that evaporation of water from wet feathers doesn't draw heat from the body of the cormorant. ✓ a

(e) Study of blood flow to different organs would be of value. ✓ It may be possible that vasodilation of arteries to the muscles allows greater blood flow and therefore more oxygen to be carried to the muscle for movement in the water. ✓ a

---

e **8/8 marks awarded** a Excellent answers throughout, for full marks.

Student B

(a) Both share a similar habitat, but the shag captures fish from the upper layers of the water ✓, while the cormorant feeds on fish at the bottom, such as flatfish. ✓ a

(b) The beak is curved at the end for gripping fish. ✓ b

(c) Waterlogging of feathers will make the cormorant heavier. ✗ c

(d) Allowing the feathers to dry after hunting means that the cormorant will be lighter and able to fly more efficiently. ✓ d

(e) It may be that the cormorant's haemoglobin has a greater affinity for oxygen. ✓ e

---

ⓔ 5/8 marks awarded **ⓐ** The descriptions of the niches are correct but no attempt has been made to explain the advantage of the differences. 2 marks awarded. **ⓑ** Correct, for 1 mark. **ⓒ** Again, there is no attempt to provide an explanation, for example that being heavier allows them to dive more effectively. No mark awarded. **ⓓ** This is a plausible suggestion and the mark is awarded. **ⓔ** This is a valid response and so is awarded 1 mark, but does not explain why this would be an advantage. Student B has failed to provide complete answers for three parts of the question.

Question 13 Distribution studies

(a) Sand dunes can differ in their species composition. The diagram below is a vertical section through one particular sand dune system.

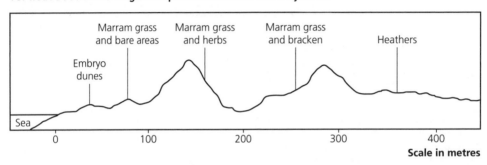

Describe how you would sample the area depicted in the diagram and use the results to illustrate quantitatively the zonation of the vegetation. (5 marks)

(b) Flowering plants were sampled, using a pin frame, in two areas of grassland near Kinnego Bay in Lough Neagh.

(i) Describe how the pin frame is used to measure plant abundance. (1 mark)

The two grasslands are named after local farmers: Wilfie's meadow is adjacent to the shoreline and is ungrazed, though cut at autumn time; Jonno's field is on slightly higher ground and is grazed only occasionally.

The results, recorded as frequencies (%), were as follows:
Wilfie's meadow: cinquefoil 5; knapweed 2; knotweed 4; loosestrife 4; meadow buttercup 4; stitchweed 2; vetchling 2
Jonno's field: cinquefoil 1; creeping buttercup 3; loosestrife 1; meadow buttercup 1; sorrel 1; speedwell 1; vetchling 1

(ii) Organise these results into an appropriate table with a suitable caption. (4 marks)

(iii) What conclusion can be made about the total diversity of the flowering plants in each grassland? (1 mark)

(iv) Apart from moisture content, what other edaphic factor might be measured? (1 mark)

Total: 12 marks

ⓔ This question assesses AO3 — understanding of *biology as an experimental science*. In part (a), you are asked about how you would undertake an ecological study of an area where there is obvious zonation. Notice that 5 marks are available, so a detailed answer is required. Part (b) involves the use of a pin frame to determine plant abundance in each of two areas, and you are asked to construct a table of the results. Make sure that you include all the information needed to allow a full interpretation.

Student A

(a) You would start at the embryo dunes and randomly sample the area ✓ using quadrats ✓. You would determine the percentage cover of each plant species. ✓ You would then move to the next area and continue until the landward area is reached. ✓ You would record the results. a

(b) (i) Pins are lowered and the plants hit are recorded. ✓ For any species the frequency is the number of hits divided by the total hits + misses. b

(ii) The results of the investigation ✗

Flowering plant species	Wilfie's meadow	Jonno's field	
Cinquefoil	5	1	
Creeping buttercup	0	3	
Knapweed	2	0	
Knotweed	4	0	
Loosestrife	4	1	
Meadow buttercup	4	1	
Sorrel	0	1	
Speedwell	0	1	
Stitchweed	2	0	
Vetchling	2	1	✓✓ c

(iii) While the species richness is the same in both fields, 6 out of 7 of the flowering plants in Jonno's field have a frequency of only 1%. Wilfie's field has a greater abundance of flowering plants and so greater diversity, possibly because it is not grazed. ✓ d

(iv) Soil pH would be an appropriate factor to measure. ✓ e

---

e 9/12 marks awarded a Student A has provided the basis of a sampling procedure: random sampling; sampling at sites transecting the dunes; use of quadrats; and determination of percentage cover to measure abundance. However, there is no attempt to suggest how the zonation would be illustrated — for example, by a series of kite diagrams (one for each species) with percentage cover (y-axis) plotted against position along the sand dunes. 4 marks awarded. b This is correct, for 1 mark. c The appropriate numbers of rows and columns have been presented, with suitable column headings, and the data correctly inserted, for 2 marks. However, the caption is not explanatory, while the measurement (frequency/%) is omitted. This should be a simple procedure, but all information should be included to allow for a full interpretation. d,e Correct.

Student B

(a) Set a line transect through the dunes and place quadrats ✓ along this at 5 metre intervals ✓. Calculate the total percentage cover of all the plants in the quadrat. ✗ Then put the results in a table and compare the different percentage covers of different zones within the sand dunes. ✗

(b) (i) The pin-frame is placed at random positions within the grassland and plants recorded. ✗

(ii) The abundance of some species of flowering plants in two areas of grassland in the vicinity of Kinnego Bay ✓

Flowering plant species	Frequency of plant species/%	
	Wilfie's meadow	Jonno's field
Cinquefoil	5	1
Creeping buttercup	0	3
Knapweed	2	0
Knotweed	4	0
Loosestrife	4	1
Meadow buttercup	4	1
Sorrel	0	1
Speedwell	0	1
Stitchweed	2	0
Vetchling	2	1

✓✓ c

(iii) There is a greater diversity of wild flowers in Wilfie's meadow. ✓ d

(iv) Light intensity. ✗ e

🄴 **7/12 marks awarded** 🄰 There are only two correct points given here. Repetition of transect lines or random sampling at points along the transect is required. It is not sufficient to determine the total percentage cover of all plants — the determination must be for each species. Again, no attempt is made to suggest how the zonation would be illustrated. 🄱 The question is about the use of the pin-frame, not its positioning. No mark gained. 🄲 This has all that an appropriate table should have: fully descriptive title, measurement with unit, suitable column headings and with all data accurately inserted. 4 marks awarded. 🄳 Correct, for 1 mark. 🄴 Edaphic factors relate to the soil.

Diversity of life

Question 14 Classifying organisms

(a) The diagram below shows *Paramecium*, a single-celled organism (length 250 µm).

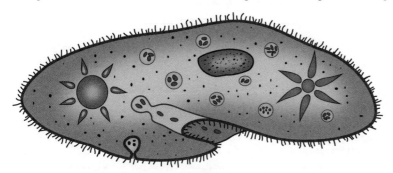

What is the classification of *Paramecium* within the:

(i) Five-kingdom system

(ii) Three-domain system (2 marks)

(b) Whales (order Cetacea) have traditionally been classified according to anatomical features. For example, one group, including the sperm whale, dolphins and porpoises, possess teeth and are classified in the sub-order Odontoceti; another group, including the humpback and fin whales, have baleen plates with which they filter small organisms from the sea, and are classified in the sub-order Mysticeti. However, modern phylogenetic relationships are based on a comparison of the base sequences in DNA and RNA.

(i) Explain the term 'phylogenetic'. (1 mark)

(ii) The base sequences in a small part of the DNA of the humpback whale and the bottlenose dolphin are presented below.

Humpback whale   TAAACCCTAATAGTCA

Bottlenose dolphin   TAAACTTAAATAATCC

How many base differences are there? (1 mark)

(iii) The relationship between five species was analysed by counting the number of common bases in a longer length of DNA (consisting in total of 52 nucleotides). The results are shown in the table below.

	Humpback whale			
Fin whale	51	Fin whale		
Sperm whale	48	48	Sperm whale		
Bottlenose dolphin	43	43	41	Bottlenose dolphin	
Harbour porpoise	45	40	45	48	

Which two species appear to be most closely related? Explain your answer. (2 marks)

(iv) What does the evidence in the table suggest about the classification of the sperm whale in the sub-order Odondoceti along with the dolphin and porpoise? Explain your answer. (3 marks)

Total: 9 marks

ⓔ This question assesses AO2. Part (a) requires careful observation of *Paramecium* structure to establish its classification. In part (b), you need to read all the information carefully. While the first part is straightforward understanding of a term, the latter parts require you to give a complete explanation *based on the evidence available*. Notice that the last part has a mark tariff of 3.

Student A

(a) (i) Kingdom Protoctista ✓

(ii) Domain Eukarya ✓ ⓐ

(b) (i) Phylogenetic refers to how closely organisms are related according to their evolutionary development. ✓

(ii) Six bases are different – 6th, 7th, 8th, 13th and 16th. ✓

(iii) The humpback and fin whales are most closely related ✓, because they have more bases in common (51) than any other pair ✓.

(iv) The sperm whale has 48 bases in common with both the humpback whale and the fin whale. ✓ It only has 41 bases in common with the bottlenose dolphin and 45 with the harbour porpoise. ✓ Because it has more common bases with the baleen whales it would appear that it is more closely related to them than to the dolphin and porpoise. ✓ The classification of whales may need to be revised. ⓐ

ⓔ **9/9 marks awarded** Student A scores full marks.

Student B

(a) (i) Prokaryotes ✗ ⓐ

(ii) Eukaryotes ✓ ⓑ

(b) (i) Phylogenetic means how closely related the organisms are. ✗ ⓒ

(ii) 5 ✗ ⓓ

(iii) The humpback and fin whales are the most closely related ✓, because they share 51 bases — more than any other two species ✓. ⓔ

(iv) The evidence suggests that the sperm whale should not be classified so closely with the dolphin and the porpoise. ✓ ⓕ

ⓔ 4/9 marks awarded ⓐ This cannot be a prokaryote since it appears to have membrane-bound organelles and is too large (250 µm). No mark gained. ⓑ This is allowed, though the correct term is Eukarya. 1 mark awarded. ⓒ Student B has not specified that phylogenetic refers to *evolutionary* relationships. No mark gained. ⓓ This was a simple count, but one base difference has been missed. ⓔ Correct answers, for 2 marks. ⓕ The statement is correct but the student has failed to provide the supporting evidence, so only 1 mark is awarded.

Question 15 Biodiversity and Simpson's diversity index

Most of the plantation forest in Ireland is conifer, mainly Sitka spruce (*Picea sitchensis*). A Sitka spruce forest was surveyed for nesting bird species. The numbers of birds of each of four species found are shown in the table below.

Nesting bird species	Number of individuals
Goldcrest (*Regulus regulus*)	8
Siskin (*Carduelis spinus*)	8
Sparrowhawk (*Accipiter nisus*)	2
Treecreeper (*Certhia familiaris*)	4

(a) Calculate the value for Simpson's diversity index (D) for nesting bird species in the survey of the Sitka spruce forest. The formula for D is:

$$D = \frac{\Sigma n_i(n_i - 1)}{N(N - 1)}$$

where:

Σ = the sum of

n_i = the total number of organisms of each individual species

N = the total number of organisms of all species

(Show your working.)   (3 marks)

In a mixed forest of Sitka spruce and ash (*Fraxinus excelsior*) a survey of nesting bird species indicated that there were 18 species present, giving a Simpson's diversity index (D) of 0.06.

(b) Use this information to explain the term 'species richness'.   (1 mark)

(c) Explain the difference between the values of Simpson's diversity index (D) for the Sitka spruce forest and the mixed Sitka spruce/ash forest. Both forests were of comparable size.   (2 marks)

(d) What do the results suggest about the strategy for planting future forests?   (2 marks)

(e) Sitka spruce is an 'introduced species'. Suggest two problems that might be associated with introduced species.   (2 marks)

Total: 10 marks

ℯ You are given the formula to use in the calculation and told what the symbols n_i and N represent, so (a) is a matter of using the numbers correctly. Don't forget that even if your answer is wrong, you can gain marks if your working is correct. This is why you must *show your working*. Parts (b) and (c) test your understanding of measures of biodiversity. In parts (d) and (e) you are asked to *suggest* — you are not expected to *know* but to provide answers that seem plausible.

Questions & Answers

(a) $D = \dfrac{8(8-1) + 8(8-1) + 2(2-1) + 4(4-1)\ \checkmark}{22(22-1)\ \checkmark} = 0.27\ \checkmark$ ⓐ

(b) This is the number of species in one area. ✓ ⓐ

(c) The mixed forest has a greater biodiversity than the Sitka spruce forest. ✓ A greater number of plant species is present, so there is a greater variety of possible nesting sites and food available for different bird species. ✓ ⓐ

(d) It is preferable to have forests planted with a mixture of tree species. ✓ This would increase the number of available niches for a variety of animal species, especially native species. ✓ ⓐ

(e) Few organisms can feed on them so the variety of organisms in the area is reduced. ✓ New diseases are introduced, so native species may be wiped out. ✓ ⓐ

ⓔ **10/10 marks awarded** ⓐ All correct, for full marks.

(a) $D = \dfrac{8(8-1) + 8(8-1) + 2(2-1) + 4(4-1)\ \checkmark}{20(20-1)\ ✗} = \dfrac{126}{380} = 0.33\ \checkmark$ ⓐ

(b) There are a lot of different species. ✓ ⓑ

(c) The mixed woodland has a smaller value for Simpson's diversity index (D) and so greater biodiversity. ✓ It is not possible to comment on why diversity is greater. ✗ ⓒ

(d) Forests should be planted according to their commercial value. ✗ ⓓ

(e) Out-competing native species. ✓ ⓔ

ⓔ **5/10 marks awarded** ⓐ The numerator part of the calculation is correct. Student B has made a mistake in totalling the numbers column, which should be 22, not 20. However, the subsequent part is not penalised, because the arithmetic operation is correct and this can be seen clearly as the working out is shown. The mistake was a slip that would have been found if Student B had checked the arithmetic. This scores 2 marks. ⓑ This is correct, for 1 mark. ⓒ The first statement is correct but Student B fails to attempt a suggestion for the greater biodiversity. Only 1 mark scored. ⓓ This is not relevant and fails to score. The question asks for a consideration of the results, i.e. the increased diversity of mixed woodland. ⓔ This is a valid point, for 1 mark. Other appropriate responses include: lack of natural predators and so a lack of control; lack of association with native species and so a threat to biodiversity; elimination of native species by, for example, predation; hybridisation with native species.

Human impact on biodiversity
Question 16 Yellowhammers and hedges

The yellowhammer, *Emberiza citrinella*, is a member of the bunting family. It feeds on grain, weed seed and grass seed in winter, though in summer adults and chicks feed mainly on invertebrates.

(a) The table below shows the classification of *Emberiza citrinella*. Complete the table by inserting the names of the three missing taxonomic groups. (2 marks)

Kingdom	Animalia
	Chordata
Class	Aves
	Passeriformes
Family	Emberizidae
	Emberiza
Species	*E. citrinella*

(b) In Northern Ireland the yellowhammer has declined by 65% over the past 30 years, and is the subject of a Northern Ireland Species Action Plan. Recommendations to reverse this decline include the following:

- Limit the use of herbicides.
- Reduce the use of insecticides.
- Restore areas of rough grass left undisturbed at field margins.

Explain how each of these recommendations might aid the recovery of the yellowhammer population in Northern Ireland. (3 marks)

(c) As we have so little native woodland in Northern Ireland, hedges are an important substitute for woodland edge habitat. As such they host a wide range of nesting birds. Different species of bird have different preferences with respect to the structure of the hedge, as shown in the table below.

Species	Preferred height of nesting site in hedge	Preferred hedge width (if any)
Yellowhammer	Low (< 1 metre)	Wide (> 2 metres)
Bullfinch	Very high (> 4 metres)	Wide
Chaffinch	High	
Goldfinch	High	Wide
Greenfinch	High	Wide
Linnet	Low (< 1 metre)	

Use this information to explain why it would be preferable to trim farmland hedges on a 3-yearly rotational basis, rather than annually. (2 marks)

(d) Apart from providing nesting sites for birds, suggest two other benefits of hedges to the biodiversity of Northern Ireland. (2 marks)

Total: 9 marks

Questions & Answers

ⓔ In part (a), notice that you need to provide three taxonomic groups for the 2 marks available. In part (b), you are required to use the information supplied in the question, together with your understanding of biodiversity, to explain aspects of a Species Action Plan for the recovery of yellowhammer numbers. Part (c) asks you to evaluate information in the question and demonstrate understanding of hedgerow management, aspects of which are also tested in part (d).

Student A

(a) phylum (✓), order ✓, genus ✓ ⓐ

(b) Herbicides kill weeds and so the yellowhammers would have no weed seeds to feed on in the winter. ✓ Insecticides kill insects, which yellowhammers feed on during summer. ✓ Restoring areas of rough grassland provides a sanctuary for grass, weeds and insects, which are important food sources for the yellowhammer. ✓ ⓑ

(c) Many species prefer high nesting sites and wide hedges and trimming the hedges every 3 years allows time for these to grow. ✓ Short hedges do not suit four of the six species. ✗ ⓒ

(d) Many hedgerow plants produce berries that are food for a variety of animals. ✓ They represent wildlife corridors that allow animals to move safely between areas. ✓ ⓓ

ⓔ **8/9 marks awarded** ⓐ All correct, for 2 marks. ⓑ This excellent answer earns all 3 marks. ⓒ High, wide hedges are preferred by some species and so 1 mark is awarded. However, some species prefer low hedges. ⓓ This excellent response scores both marks.

Student B

(a) phylum, (✓) division, ✗ genus ✓ ⓐ

(b) Yellowhammers eat weed seeds in the winter and these may be covered in herbicides, which could poison the yellowhammers. Insecticides may also be toxic. ✓ Areas of rough grass left undisturbed would be free of herbicides and insecticides. ✗ ⓑ

(c) This would allow the hedges to grow upwards and outwards, so providing birds with their preferred nesting sites. ✓ Cutting annually may kill the hedges. ✗ ⓒ

(d) Hedges provide shelter for cattle. ✗ They provide safe pathways for animals to move without attracting the attention of predators. ✓ ⓓ

ⓔ **4/9 marks awarded** ⓐ Phylum and genus are correct, for 1 mark. Division is another name (mostly used within the plant kingdom) for a phylum. ⓑ 1 mark has been allowed for the possible toxic effect of pesticides (herbicides and insecticides). Student B fails to use the information provided in the question stem: that the yellowhammer relies on weed seed (which herbicides would remove) during the winter, and on insects (killed by insecticides) as food during the summer, while areas of undisturbed rough grass would represent a source of both seeds and insects throughout the year. **c** High, wide hedges are preferred by some species and so 1 mark is awarded. However, some bird species prefer low hedges. In essence, a 3-year rotational trimming regime should mean that, at any one time, some of the hedges (i.e. those recently cut) are low while some (those uncut) would be high and wide, so there would be a variety of hedge types to suit different species. **d** The first answer, while a correct statement, does not provide a benefit to 'biodiversity', as asked for in the question. The second is correct and earns 1 mark.

Question 17 Agricultural management and pollution

There are a number of ways in which farms are managed to reduce the risk of pollution:

- Organic waste, such as manure and slurry, is stored in a secure location to prevent seepage, and allowed to decompose.
- Organic waste may be used in an anaerobic digester to produce a methane-rich biogas, which is burned to generate heat and electricity.
- Manure that has decomposed for some time is spread onto agricultural land.
- Organic waste and manufactured fertilisers are only spread at certain times of the year and then not during certain weather conditions.
- Soil is analysed and the results used in preparing a suitable fertiliser for the intended crop.
- Stubble and crop residues are not burned but ploughed into the soil before establishing the next crop.

(a) Explain the consequence of organic effluent leaking from its storage facility. (2 marks)

(b) Decomposition of organic waste is partly aerobic, producing carbon dioxide, and partly anaerobic, producing methane. Explain why it would be preferable that decomposition was aerobic rather than anaerobic. (1 mark)

(c) Suggest how the use of anaerobic digesters to generate electricity should lessen the impact on global warming. (1 mark)

(d) Suggest how spreading rotted manure might reduce pollution. (1 mark)

(e) Under which weather conditions is it not permissible to spread organic or manufactured fertiliser? Explain why. (1 mark)

(f) Suggest how analysing soil so that only an appropriately prepared fertiliser is used reduces the risk of pollution. (1 mark)

(g) Suggest why stubble and crop residues are not burned. (1 mark)

Total: 8 marks

e The question requires you to read the information about farmland management and then *apply your understanding* (AO2). Ensure that your answers relate to pollution — atmospheric (and its effect on global warming) or of waterways.

Student A

(a) If there was leakage of organic effluent it could run off into neighbouring waterways, where it would be fed on by aerobic bacteria. ✓ The subsequent population explosion of bacteria would deplete the water of oxygen, causing the death of many aquatic animals. ✓ ⓐ

(b) Methane, produced by anaerobic decomposition, is 20 plus times more effective as a greenhouse gas than CO_2. ✓ ⓐ

(c) Electricity produced by anaerobic digesters replaces electricity produced from fossil fuels and so reduces CO_2 emissions. ✓ ⓐ

(d) It reduces the use of manufactured fertilisers, which are implicated in the release of nitrous oxide, the most potent greenhouse gas, into the atmosphere. ✓ ⓐ

(e) Not allowed during periods of rain, to reduce the risk of runoff into waterways. ✓ ⓐ

(f) Prevents surplus amounts of fertiliser being spread, which would increase the release of nitrous oxide into the atmosphere. ✓ ⓐ

(g) To prevent CO_2, a greenhouse gas, being released into the atmosphere. ✓ ⓐ

ⓔ **8/8 marks awarded** ⓐ Excellent answers throughout, for full marks.

Student B

(a) To prevent runoff into waterways. ✗ ⓐ

(b) CO_2 produced by aerobic decomposition, is a much less potent greenhouse gas than methane. ✓ ⓑ

(c) The use of the digester and burning of the biogas produced reduces the release of methane, a potent greenhouse gas, into the atmosphere. ✓ ⓒ

(d) It reduces the need for artificial fertiliser and, since its manufacture requires energy, it reduces the use of fossil fuels. ✓ ⓓ

(e) Not spread during rainy periods as runoff into streams is likely. ✓ ⓔ

(f) If the amount of fertiliser applied was not the same as that used by the crop then some may run off into rivers, causing eutrophication. ✓ ⓕ

(g) Stubble provides seed for overwintering birds. ✗ ⓖ

ⓔ **5/8 marks awarded** ⓐ This lacks detail and scores no marks. Student B has failed to explain the consequence of runoff into waterways. ⓑ Correct, for 1 mark. ⓒ An alternative though valid answer, for 1 mark. ⓓ Another alternative to that provided by Student A. 1 mark awarded. ⓔ Correct, for 1 mark. ⓕ Another alternative answer for 1 mark. ⓖ This is not relevant to the context of the question, which is about pollution.

■ Section B Essay questions

Question 18 The movement of water through the plant

Give an account of the processes involved in the movement of water through a plant. (15 marks)

In this question you will be assessed on your written communication skills.

Total: 15 marks

ⓔ In answering this question (worth 15 marks) you must write in continuous prose because your quality of written communication (QWC) will be assessed. This is a big topic, so take some time to *devise a plan*. Be aware that 'the movement of water through a plant' involves what happens in the root, in the stem and in the leaf. You must be able to sequence your ideas and use the appropriate biological terms, which is how QWC will be assessed.

Student A

Water enters the root and moves through the cortex via the apoplast or symplast pathways. ✓ Using the apoplast route, water moves through the cellulose cell walls. ✓ This is the main route ✓ since there is less resistance to movement. Water moves through the cytoplasm of cells and from cell to cell via plasmodesmata ✓ using the symplast route ✓. Water can only move through the endodermis, and into the xylem, using the symplast route. ✓

Forces of adhesion between the water molecules and the lignified xylem vessels ✓ help the water to creep up the xylem ✓. The force of cohesion, due to the attraction between water molecules, maintains a continuous water column. ✓ This water column moves up as water leaves the xylem to replace ✓ the water that has evaporated from the mesophyll surface ✓ and diffused out of the open stomata ✓. Throughout, water is moving along a water potential gradient ✓ as transpiration creates a particularly negative water potential in the leaf ✓. ⓐ

Quality of written communication ⓑ

ⓔ 15/15 marks awarded ⓐ Student A has 14 correct points in the content. ⓑ This is a well-structured account and ideas are expressed fluently. The links within the overall process are made clearly, all points are sequenced and biological terms are used accurately. Since the quality of written communication is of a very high standard, a total of 15 marks is judged to be appropriate.

Student B

In the root, water travels from cell to cell via two routes: the symplast and the apoplast. ✓ Apoplast is when the water travels through each cell✗, whereas symplast is when water travels along the cell walls. ✗ However, water cannot pass through the Casparian strip ✓ and so must enter the cell before passing into the xylem. ✓ Water is then drawn upwards by negative tension, adhesion and cohesion. Negative tension is the force produced by water evaporating out of the leaves. ✗ Adhesion is when the polar water molecules ✓ are attracted to the sides of the xylem ✓. Cohesion is when water molecules are attracted to each

---

other. ✓ Once the water travels up the stem it enters the leaves. The water is then passed from cell to cell and once it reaches the outer layer of cells next to the air space it evaporates ✓ and is released through the stomata. a

Quality of written communication b

ⓔ **9/15 marks awarded** a Seven appropriate points are provided. Some statements are simply wrong — for example, the symplast and apoplast routes have been confused. Other statements are not sufficiently precise: water does not evaporate out of the leaf; water vapour diffuses out following evaporation from the mesophyll surface. A number of relevant points (for example reference to the endodermis in the root) have been omitted. You should read Student A's answer for a fuller response. b Student B expresses ideas clearly, even though there are some factual errors, and the account is reasonably well sequenced. The use of specialist vocabulary is generally appropriate. As the quality of written communication is of a high standard, a total of 9 marks are awarded.

Question 19 The cardiac cycle

Give an account of the phases of diastole, atrial systole and ventricular systole during the cardiac cycle. The account should make reference to each of the following:
- **the waves of excitation**
- **the pressure changes**
- **the opening and closure of valves** (15 marks)

In this question you will be assessed on your written communication skills.

Total: 15 marks

ⓔ This is quite an involved essay and a plan will help you. As you describe the events of the phases of the cardiac cycle — diastole, atrial systole and ventricular systole — you must refer in each case to the wave of excitation over the cardiac muscle, the contraction (and relaxation) of atria and ventricles, the changes in pressure within the heart chambers and associated arteries and how pressure differences cause opening and closing of atrioventricular and semilunar valves, and the subsequent movement of blood. Do you see the benefit of a plan?

Student A

Diastole: During diastole the heart is relaxed. ✗ a

Atrial systole: During atrial systole an electrical impulse is generated within the sinoatrial node. ✓ This wave of excitation propagates over the walls of the atria and causes contraction. ✓ This results in the remaining blood within the atria being pushed into the ventricles. ✓ Blood flows down the pressure gradient and the atrioventricular valves are pushed open. b

Ventricular systole: The wave of excitation reaches the atrioventricular node. ✓ The electrical impulses then spread downwards through the bundle of His ✓ and the Purkinje fibres, stimulating contraction of both ventricles from the apex upwards and pushing blood towards the arteries ✓. As the pressure in

---

the ventricles increases, the atrioventricular valves are slammed shut. ✓ As the pressure increases further it becomes greater than in the arteries and the semilunar valves are forced open. ✓ c .

Diastole: When the atria are relaxed blood is returned via the venae cavae (right atrium) and the pulmonary veins (left atrium). ✓ As blood returns, pressure increases and, when the ventricles relax, pressure in the atria increases to a point above that in the ventricles, so that the atrioventricular valves are pushed open. ✓ Both ventricles then fill passively. ✓ d

Quality of written communication e

---

e **13/15 marks awarded** a This is not sufficient. However, some events of diastole are discussed appropriately later in the answer. b There are three correct points here, although some detail is missing. The AV valves are already open during atrial systole, because the atrial pressure increases when blood is returned to the heart. c This is a reasonable account of ventricular systole though, again, some detail is missing — for example, there is no mention of the role of the chordae tendinae in preventing the AV valves turning inside out. Student A has 5 correct points. d This is a good account, even if some detail has been omitted. Student A has 3 points here, giving a total of 11 points for the biological content of the answer. e This is a well-structured account and ideas are expressed fluently using appropriate specialist language. The links between electrical activity and contraction, and between pressure changes and valve action, are made clearly. Since the quality of written communication is of a high standard, a total of 13 marks is judged to be appropriate.

Student B

Diastole: This phase of the cardiac cycle is the resting phase. On the right side of the heart blood flows into the right atrium and on the left side it flows into the left atrium. ✗ a

Atrial systole: During atrial systole the right atrium fills with blood from the venae cavae and the left atrium from the pulmonary veins. ✗ The atria contract and increase pressure in the atria forcing blood into the ventricles ✓, because when the pressure in the atria is greater than in the ventricles the atrioventricular valves open ✓. Atrial systole begins with electrical activity being produced at the sinoatrial node. ✓ b

Ventricular systole: During ventricular systole, the tricuspid and bicuspid valves are forced shut ✓ and prevent backflow into the atria. The semilunar valves are forced open. ✓ This allows oxygenated blood from the left ventricle to enter the aorta, from where it will supply the organs of the body; blood from the right ventricle enters the pulmonary artery and is taken to the lungs. ✓ The semilunar valves slam shut and the pocket-like valves prevent backflow to the ventricles. ✗ c

Quality of written communication d

Questions & Answers

e **7/15 marks awarded** **a** There is nothing incorrect here; it just lacks the detail necessary. For example, where is the blood flowing from (the major veins), what happens to the pressure in the atria, what effect has this on the atrioventricular valves? Student B fails to score. **b** The atria fill with blood during diastole, not during atrial systole, and so the first statement is wrong. The second statement is correct. The opening of the atrioventricular valves actually occurs during diastole, but the concept is correct and so the third statement is awarded a point. Then there is a correct reference to the SA node initiating the wave of excitation. 3 correct points are recognised. **c** Student B notes correctly that the atrioventricular valves are forced shut and that, at a later time, the semilunar valves are forced open. However, there is no reference to the increasing pressure in the ventricles, that the atrioventricular valves are only forced shut when the pressure is greater than that in the atria and that the semilunar valves are forced open when the pressure in the ventricles is greater than in the major arteries. Semilunar valves close during diastole, when the pressure in the ventricles becomes lower than in the major arteries. Overall, while there is good understanding of the structure of the heart, there is a lack of understanding of the pressure changes and the electrical activity that coordinates the contractions. Student B makes 3 points here, giving a total of 6 points for the biological content of this answer. **d** The ideas are not well sequenced, though there has been appropriate use of some specialist terms. The quality of written communication is judged to be of a reasonable standard and so a final total of 7 marks are awarded.

Knowledge check answers

1 nitrate (and sulfate) ions
2 Surface area = 22, volume = 6, so the surface area-to-volume ratio is 22:6 or 11:3.
3 Humans are terrestrial (with an impermeable body surface); they are large (small surface area-to-volume ratio); they have a high metabolic rate (large demand for oxygen).
4 They are especially small and have a biconcave disc (flattened) shape.
5 diffusion
6 A good transport system has a pump (or equivalent), vessels and a fluid to transport materials around the body. It also has exchange surfaces to load substances into the transport system and to remove them where required.
7 It increases the surface area-to-volume ratio (more oxygen is absorbed relative to the amount required) *and* means that all tissues are close to the absorbing surface.
8 Plants lack muscle tissue, i.e. they do not move from place to place.
9 Leaves have an air space system; they are thin (and broad); they have stomata that connect to the atmosphere.
10 They allow air to enter from the atmosphere: carbon dioxide for photosynthesis; and oxygen, which may diffuse down to the roots through an extensive air-space system, for respiration. Both oxygen and carbon dioxide are more plentiful in air than dissolved in water.
11 Ventilation of the lungs replenishes oxygen-rich air in the alveoli, thus increasing the oxygen diffusion gradient into the blood. It also replaces air that has a high carbon dioxide concentration with air that has a low carbon dioxide concentration, thus increasing the carbon dioxide diffusion gradient out of the blood.
12 Inhalation depends on the contraction of muscles of the diaphragm and the external intercostal muscles of the ribcage, while exhalation normally involves only the relaxation of these muscles.
13 The alveoli collapse and are difficult to expand, making breathing difficult and reducing the efficiency of gas exchange. Blood oxygen levels fall and carbon dioxide levels increase.
14 Emphysema reduces the efficiency of breathing (reduced recoil of alveoli due to loss of elasticity) and gas exchange (reduced surface area of alveoli). The overall result is that oxygen uptake is reduced greatly, so that less is available for the production of ATP in respiration — hence the difficulty in undertaking strenuous activity.
15 a Xylem is the star-shaped tissue located centrally in the root and on the inner sides of the vascular bundles in the stem.
   b Phloem is found between the arms of the xylem in the root and on the outer sides of the vascular bundles in the stem.

16 Xylem vessels lack cross walls, lack any cytoplasm (to impede the flow of water), are lignified (to keep them open and prevent them collapsing) and have pores (to allow water to leave laterally).

17 Companion cells contain numerous mitochondria that produce the ATP used to carry out active processes such as loading sucrose etc. into the sieve tubes.

18 Lack of oxygen reduces the rate of aerobic respiration, which produces the ATP used in the active uptake of ions.

19 Root pressure results from the active secretion of ions into the xylem. A respiratory poison prevents the production of ATP used in active transport.

20 Water vapour diffusing out of the stomata is trapped and the region immediately outside becomes saturated. The water potential gradient between the inside of the leaf and the outside is reduced.

21 Very young leaves are removing sucrose from the phloem (they are a 'sink') while mature leaves are releasing sucrose since they are photosynthesising at a high rate (they are a 'source').

22 Substances enter the blood at the exchange (absorptive) surface. These substances have to be transported quickly enough to satisfy the organism's requirements (which will be great if it is large and active). Diffusion alone is not rapid enough.

23 A double circulation allows a two-pressure system with higher pressure to the body cells and lower to the lungs. High pressure in the systemic (body) circulation ensures that oxygenated blood is pumped efficiently to the body cells and allows the formation of tissue fluid. Low pressure in the pulmonary (lung) circulation ensures that fluid is not forced out into the alveoli and the slower flow of blood allows more time for gaseous exchange.

24 vena cava → right atrium → right ventricle → pulmonary artery → capillaries in a lung → pulmonary vein → left atrium → left ventricle → aorta

25 They have the same thickness of muscle and contract with equal force, since they only have to pump blood into the ventricles below.

26 They close when the pressure in the ventricles is greater than the pressure in the atria.

27 They are closed during diastole and atrial systole when the ventricle muscle is relaxed. They are also closed during the initial part of ventricular systole when the ventricle muscle contracts without raising the ventricular pressure above that in the artery. They are closed as long as the arterial pressure exceeds that in the ventricles.

28 $\dfrac{60\,\text{s}}{0.8\,\text{s}}$ = 75 beats min^{-1}

29 The SAN initiates the heartbeat by emitting impulses over the atria. The AVN delays these impulses and relays them to the bundle of His, which conducts them to the base of the ventricles (to the Purkinje fibres).

30 a The delay allows time for the atria to empty before the ventricles contract.

b By contracting from the base upwards, each ventricle forces all the blood towards the arteries.

31 The aorta, because the aortic pressure is much greater than that in the pulmonary artery.

32 They pick up oxygen in the lungs and release it in respiring tissues.

33 It has no nucleus, so has more room for haemoglobin (each red blood cell contains around 250 million haemoglobin molecules). It is small and flattened, so has a large surface area-to-volume ratio. Being thin means that oxygen does not have to diffuse far to reach a haemoglobin molecule.

34 Polymorphs and monocytes (which develop into macrophages). Polymorphs are short-lived, while monocytes are long-lived.

35 Sucrose is transported in plants; glucose is transported in animals.

36 Thromboplastin (thrombokinase) converts prothrombin (a blood protein) to thrombin (another enzyme) which converts fibrinogen (another blood protein) to fibrin (which forms the fibres of the clot).

37 There are two factors — actively respiring tissues consume more oxygen and release more carbon dioxide. Both a decrease in pO_2 and an increase in pCO_2 cause more oxygen to be released from the haemoglobin.

38 At the pO_2 found in the placenta, fetal haemoglobin must be able to pick up oxygen when the maternal haemoglobin is releasing oxygen.

39 Benefit: if you know you are at high risk of cardiovascular disease, you can take action to reduce other risk factors. Potential problem: insurance companies might want to know if you are at high or low risk, which would affect the premiums that you pay. (Other suggestions are valid.)

40 A feature (behavioural, physiological, etc.) that helps an organism survive in its habitat, breed and pass on its genes.

41 Edaphic factors are environmental factors that relate to the soil.

42 Test whether they can interbreed successfully and produce fertile offspring.

43 The Latin names are used internationally and are understood by scientists worldwide, whereas common names can differ locally let alone internationally.

44 Plants are autotrophic (photosynthetic), containing chloroplasts with chlorophyll, while fungi are heterotrophic (lysotrophic). Plants are multicellular, while fungi may be unicellular, though most are multicellular. Plants store carbohydrate as starch, while fungi store carbohydrate as glycogen. Plant cells have a cellulose cell wall, while fungi have cell walls made of chitin.

Knowledge check answers

45 a Chitin cell wall/cells are organised into hyphae.
   b Animal cells never have a cell wall; they store lipids as fats.

46 Both are heterotrophic and store carbohydrate as glycogen.

47 The two species are not closely related. Separation at low temperature indicates that there are few hydrogen bonds holding the two strands together. This means that there are many bases in the hybrid DNA that are not complementary — for example if the bases C and A come together during hybridisation, they will not form a hydrogen bond.

48 Seed banks maintain a source of genetic diversity for crop plants. They contain genes that could confer subtle variations, such as resistance to drought, frost or pests (which modern plants may lack, having been selected for increased yield).

49 Advantages: reduced need for fertiliser (because soil fertility is maintained) and pesticides (because build-up of pests is prevented). Disadvantages: reduced profitability through higher labour costs and greater diversity of machinery required.

50 Only a few species are adapted to a habitat of short, well-trimmed hedges. A variety of hedge heights and thicknesses, resulting from a 3-year rotation of hedge trimming, will favour a greater biodiversity because some species will be adapted to conditions in tall, wide hedges while others will be adapted to short, narrow hedges.

51 Simpson's diversity index, D, would be lower for the mixed woodland than for the conifer plantation.

52 This is because of pest resistance, pest resurgence and secondary pest outbreak.

53 The high levels of oxygen released support a large bacterial population capable of digesting organic matter in the discharge, as long as the latter is not excessive.

54 a Higher temperatures stimulate algal growth (increasing enzyme activity and algal metabolism).
   b Movement of the water (e.g. tumbling over rocks) re-oxygenates the water.

55 Organic matter needs to be broken down — by soil bacteria and fungi — to release nutrients, and this takes place over time.

56 To protect any individual species (such as an orchid) it is necessary to protect its habitat; and, more importantly, it is likely to be the whole community of plants and animals in the habitat that needs protection.

57 species diversity, genetic diversity within species, ecosystem diversity

58 It leads to nutrient enrichment of the soil, which selects rapid-growth grass species. These outcompete plant species that are adapted to nutrient-poor soils, which therefore results in a loss of biodiversity.

59 As permafrost thaws, bacteria within the soil will become active, feed on the organic matter and through their respiration release large amounts of CO_2 and methane into the atmosphere. This will result in a projected additional global warming of 0.13–1.69°C by 2020.

60 Some of these mosquitoes are vectors for a number of tropical diseases, including malaria.

61 The rise in carbon dioxide levels is caused by combustion of fossil fuels and deforestation.

Index

Index